Black & Decker
POWER TOOL
CARPENTRY

VNR VAN NOSTRAND REINHOLD COMPANY
NEW YORK   CINCINNATI   TORONTO   LONDON   MELBOURNE

Consultant editors: George Daniels and Tom Philbin

Copyright © 1978 by Marshall Cavendish Limited
Library of Congress Catalog Card Number 77-90840
ISBN 0-442-01985-8

Printed in Great Britain.

Published in 1978 by Van Nostrand Reinhold Company
A Division of Litton Educational Publishing, Inc.
450 West 33rd Street, New York, N.Y. 10001, U.S.A.

This publication is not to be sold outside the United States
of America and its territories.

1   3   5   7   9   11   13   15   16   14   12   10   8   6   4   2

Library of Congress Cataloging in Publication Data
Main entry under title:

Black and Decker power tool carpentry.

   1. Carpentry – Tools.   2. Power tools.   3. Black &
Decker Manufacturing Company, Towson, Md.   1. Daniels,
George Emery, 1914-   II. Philbin, Thomas, 1934-
III. Title: Power tool carpentry.
TH5618.B53 1978   684'.083   77-90840
ISBN 0-442-01985-8

INTRODUCTION

Trying your hand at home carpentry is something of a gamble,
but *power-tool* carpentry virtually guarantees you success –
even with the most exciting and ambitious of projects.
This book has been specially written to enable *anyone* to quickly
master the basic power-tool techniques and put them into
practice on projects chosen for their qualities
of design and ease of construction.
In fact, the only difficulty you are likely to come across is
deciding which project to start first, since the choice includes
virtually everything from a garden gate to a 4-poster bed!

CONTENTS

Publisher's note

The lumber for the projects in this book has been selected for its relatively easy availability. If you have difficulty obtaining a particular size stock, it is suggested that you buy the next largest size and have it trimmed down by the lumberyard (they should do this for a nominal fee) or trim it yourself.

Techniques

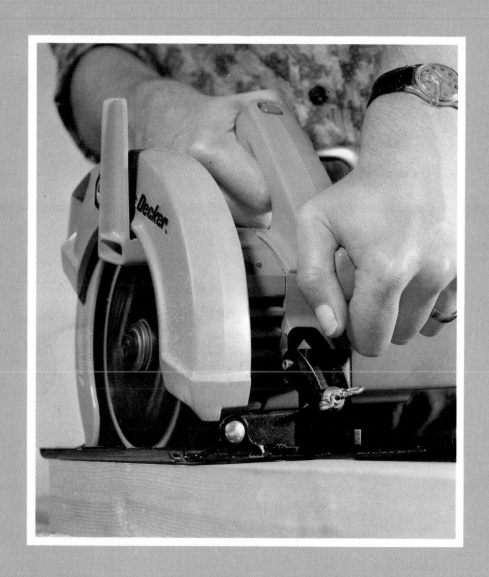

Basic tool kit

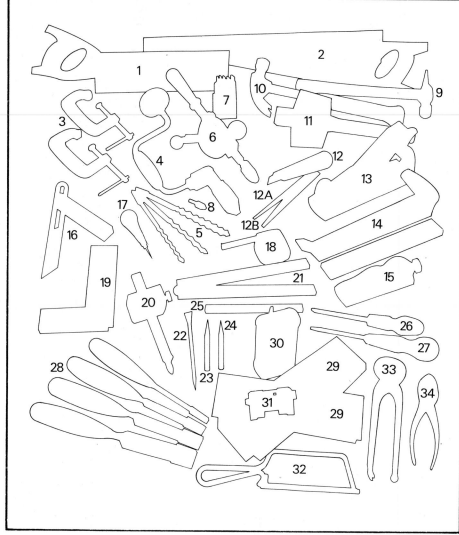

The Tools for the Job

One of the dullest and slowest jobs in carpentry is making a large number of identical pieces—sawing the same shape repeatedly, hand-drilling row after row of holes, sanding to shape and so on. Power tools take the routine and heavy labor out of many of these tasks so that your enthusiasm and energy are not diminished and your work will be of a better standard. Power tools will also enable you to complete a better looking product, because you can achieve greater accuracy of line not only when sawing long edges but also when making strong, durable joints.

However, the power tool kit will need to be complemented by the carpenter's basic tool kit. This should consist of the following tools.

1. Tenon or back saw. These saws are available in blade lengths of between 8in and 14in (203mm and 356mm) with 13, 14, 15, 16 or 20 teeth per inch. This is used for jointing and cutting across the grain on small pieces. The back of the blade may be of brass or steel. The saw with 20 teeth per inch is for cutting dovetails and it has a thin blade to give greater accuracy. The dovetail saw performs a ripping action, so cut along the grain when using it.

2. Handsaw. This is used for cutting larger pieces of lumber. There are three types of handsaw. The one shown here is a panel saw. It is 20in to 22in (508mm to 559mm) long with 10 teeth per inch. Its special purpose is for fine crosscut and jointing work for and cutting plywood, blockboard and hardboard. The other

types of handsaw (not shown) are the rip saw and the crosscut saw. The rip saw is 26in (660mm) long and 5 teeth per inch. Its special purpose is for cutting softwoods, working with the grain. The crosscut saw is 24in to 26in (610mm to 660 mm) long with 6 to 8 teeth per inch and is used for cutting across the grain of hardwoods and softwoods and for working with the grain on very hard woods.

3. C clamps. These are used for a range of clamping purposes. These clamps are available in a 1in to 18in (25mm to 457 mm) range of opening and between 1in to 8in (25mm to 203mm) depth of throat. When using C clamps always place a waste scrap of lumber between the piece to be clamped and the jaws of the clamps. This prevents bruising of the piece.

4. Ratchet brace. This has spring-loaded jaws in a screw-tightened chuck. It is specially designed for holding wood auger bits (5). The brace is available with or without a reversible ratchet in a sweep (the arc described by the turning handle of the brace) ranging from $5\frac{7}{8}$in (148mm) to 14in (355mm).

5. Wood auger bits. These are used with the ratchet brace (4).

6. Hand drill. This is used for holding wood and metal twist drill bits (7) and countersink bits (8). The example shown here has a double pinion (cogged drive wheel).

7. Twist bits. These are commonly available in sizes ranging from $\frac{1}{64}$in to $\frac{1}{2}$in (0.4mm to 13mm). The type of steel used depends on the use to which the bit is to be put.

8. Countersink bit. This is used for countersinking drilled holes so that countersunk screwheads will fit flush with the surface of the piece you are working with.

9. Warrington pattern or cross peen hammer. This is used for general nailing and joinery and can be used for planishing and beating metal. Weights of these hammers range from 6oz (170g) to 16oz (450g).

10. Claw hammer. This is used for general purpose carpentry, in particular for driving and removing nails. When taking out nails, make sure that the nail head is well into the claw of the hammer and, if it is necessary to protect the surface of the wood, place a scrap piece of lumber between the claw and the wood. Exert even pressure to lever the nail out. Claw hammers are available in weights ranging from 16oz (450g) to 24oz (570g).

11. Carpenter's or joiner's mallet. This is used for general carpentry and cabinet work and is available in head lengths of between 4in (102mm) and 5½in (140mm).

12. Handyman's knife. This useful carpentry knife can be fitted with a variety of blades to suit specific purposes. The blades include angled concave, convex, linoleum and hooked blades. Wood and metal saw blades (12A and 12B) can also be fitted to this tool as can a blade for cutting plastic laminate.

13. Bench plane. There are various types of bench plane and they are available in a range of lengths and widths. The *Smooth plane* (shown here) comes in lengths of between 9½in and 10¼in (241mm and 260mm) and widths of between 1¾in and 2⅜in (45mm to 60mm). The *Jack plane* (not shown) is available in lengths of between 14in (356mm) and 15in (381mm) and widths ranging from 2in (51mm) to 2⅜in (60mm). The *Fore plane* (not shown) is 18in (457mm) long and 2⅜in (60mm) wide. The *Jointer plane* (not shown) is 22in (561mm) long and 2⅜in (60mm) wide. When working with resinous or sticky woods, a plane with a longitudinally corrugated sole makes the job of planing easier because friction between the lumber and the plane is reduced. If you do not have such a plane, apply a drop of vegetable oil to the sole of your ordinary plane—this will perform much the same function.

14. Surform plane. This is one of a range of open rasp/planing tools, all of which are useful and versatile. They are primarily used for rough work but with care some reasonably fine craftmanship can be produced. Each tool in this range has replaceable blades.

15. Block plane. This small plane is particularly useful for fine cabinet work and for planing end grain. Available in lengths of between 6in and 7in (152mm to 178mm) and cutter widths of between 1 15/16 in (49mm) and 1⅝in (41mm).

16. Sliding bevel. This tool is used for setting out angles, or bevels. Available in blade sizes of 9in (239mm), 10½in (267 mm), and 12in (305mm).

17. Bradawl or scratch awl. This is a chisel pointed boring tool used for marking screw position and counterboring for small size screws.

18. Adjustable steel rule. The pocket size variety, when fully extended, range in length from 6ft (1.83m) to 12ft (3.66m). The larger varieties are available in either steel, fiberglass or fabric in lengths of up to 100ft (30.5m).

19. Try square. This is used for setting out right angles and for testing edges when planing lumber square. The tool has a sprung steel blade and the stock is protected by a thin strip of brass or other soft metal. Available in blade lengths of 6in (152mm), 7½in (190mm), 9in (230mm) and 12in (300mm).

20. Marking gauge. This is used to mark one or more lines on a piece of lumber, parallel to one edge of that lumber. The type shown here is a mortise gauge which has a fixed point on one side and one fixed and one adjustable point on the other. Its specific use is for marking out mortise and tenon joints but it can be used in the same way as an ordinary marking gauge.

21. Folding wooden rule. This tool is also available in plastic. Primarily for joinery and carpentry use, it should be used narrow edge onto the lumber for the most accurate marking. These rules are available in 2ft (600mm) and 3ft (1m) sizes.

22. Scriber marking knife. One end of this tool is ground to a chisel shaped cutting edge for marking lumber. The other end is sharpened to a point and can be used for scribing metal.

23. Punch or nail set. This tool is used for tapping tack and nailheads below the surface of lumber. A range of head sizes is available to suit all nail sizes.

24. Center punch. This is used for spot marking metal to give a guide for drilling. The point is marked by tapping the wide end of the tool with a hammer. Automatic center punches (not shown) are available. These are spring loaded so you do not have to tap the end of the tool.

25. Carpenter's pencil. This has an oblong shaped lead which is sharpened to a chisel edge so that it can be used to black in lines scribed on lumber.

26. Phillips or Pozidrive type screwdrivers. This tip is designed for use with crosshead type screws which are increasingly replacing screws with the conventional blade head. This crosshead design allows far greater contact between the screwdriver tip and the screwhead— providing, of course, that the correct size of screwdriver tip is used. This makes for greater torque (twisting power) and

reduces the likelihood of tool slip and consequent damage to the work.

27. Screwdriver. This tool is available in blade lengths of between 3in (76mm) and 18in (457mm) and tip widths of between 3/16 in (4.8mm) and ½in (13mm). The screwdriver tip should fit the screw slot completely and the risk of tool slip will be further reduced if the screwdriver tip has been cross ground.

28. Carpenter's chisels. These are available in several shapes and sizes of both handles and blades. The firmer bevel edge chisels shown here are probably the most useful all around chisels to have in a basic tool kit. Chisel handles are either of ash, boxwood or plastic (shown here). Plastic handles are virtually unbreakable on quality chisels but wood handles should be treated with care and should only be hit with a wooden mallet. Blade widths vary from ⅛in (3mm) to 2in (51mm).

29. Oilstones. These are used for sharpening the cutting edges of such tools as planes and chisels. There are two main kinds of oilstone, natural and artificial. Natural stone comes in several types. *Washita* gives a good finish and cuts well. *Arkansas* is an expensive stone but it is of high quality and produces a very fine edge. These are the most commonly used natural oilstones. Artificial stones come in three grades—coarse, medium and fine—and have the advantage of maintaining their quality. They are available in a selection of sizes including 5in x 2in (127mm x 51mm), 6in x 2in (152mm x 51mm), 8in x 2in (203mm x 51mm), 10in x 2in (254mm x 51mm) and 8in x 1⅞in (203mm x 46mm).

30. Fine machine oil. This has many lubricating uses in the workshop and can be used in conmunction with oilstones.

31. Honing gauge. This is a useful device for holding bladed tool at the correct angle for sharpening on an oilstone. The disadvantage of this tool is that it tends to cause wear in the center of the oilstone rather than distributing the wear evenly over the whole stone.

32. Small hacksaw. This is a general purpose saw for light metalworking jobs.

33. Pincers. These are used for pulling nails and tacks from lumber. If possible, always place a scrap of waste lumber between the jaws of the pincers and the work piece to avoid bruising.

34. Slip-joint pliers. This tool has a thin section so that the jaws can reach into tight places. It has two jaw opening positions and shear type wire cutter.

1. Butt joint

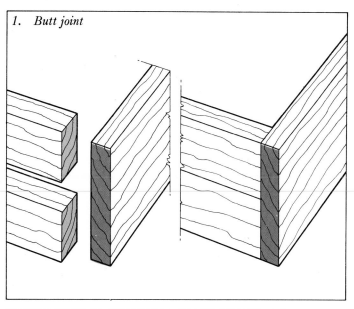

1. The butt joint is the simplest of all joints in carpentry. It may be made either straight or right-angled, and also needs nails or screws, sometimes with the addition of glue, to hold it firmly joined.

2. The dowelled joint is basically a butt joint reinforced with dowels – lengths of wooden rod. Both halves of the joint are often drilled at once to make the holes line up.

3. The secret dowelled joint is better-looking because the end of the dowels do not show. The two rows of holes are drilled separately, so great accuracy is essential.

4. The 45° mitered joint has a very neat appearance, because no end grain is visible. Unfortunately, it is a very weak type of joint unless it is reinforced in some way, for instance, with a corrugated staple.

5. The end-lap joint is used at the corners of a rectangular frame. It is simple to make, has a reasonably neat appearance, and is quite strong if glued together.

2. Dowelled joint

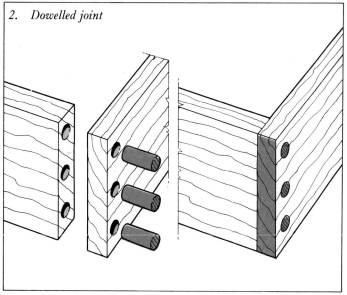

4. 45° mitered joint

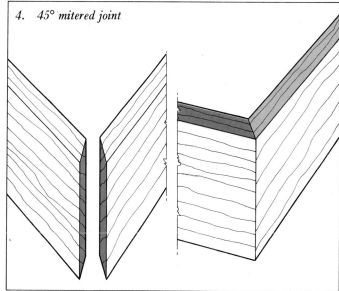

3. Secret dowelled joint

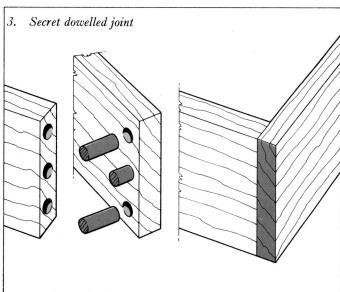

5. End-lap joint

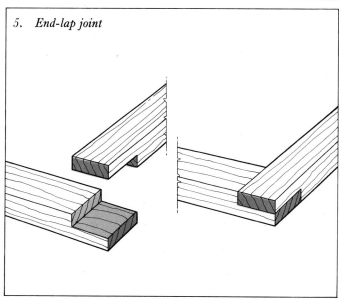

6. The middle-lap joint is a variant of the usual end-lap. It is generally used in conjunction with the previous type of lap-joint in the construction of simple frameworks, for instance.

7. The cross-lap joint is the third member of this versatile family. This type of joint should be used where two pieces of lumber have to cross without increasing the thickness of the frame.

8. The lapped-dovetail joint is an extra-strong lap joint. Its angled sides make it impossible for it to be pulled apart in a straight line, though it is important to note that it still needs glue to hold it together.

9. The housed joint is used for supporting the ends of shelves, because it resists a downward pull very well. It, too, must be reinforced with glue or screws.

10. The stopped-dado joint has a neater appearance, but is harder to make because of the difficulty of cutting out the bottom of the rectangular slot neatly.

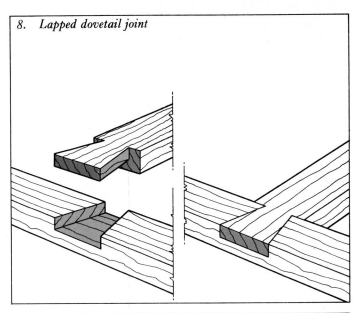

8. *Lapped dovetail joint*

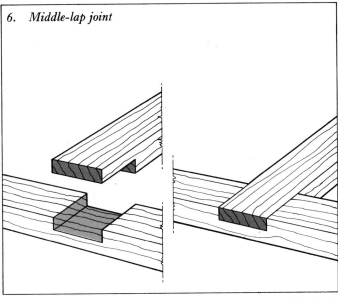

6. *Middle-lap joint*

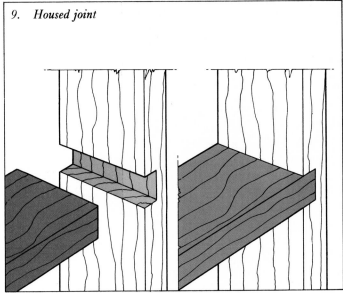

9. *Housed joint*

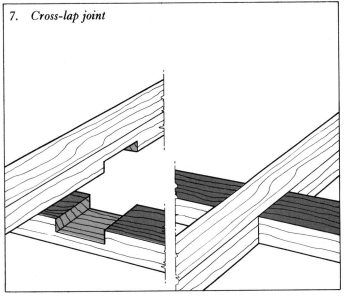

7. *Cross-lap joint*

10. *Stopped-dado joint*

11. Tongue-and-groove joint

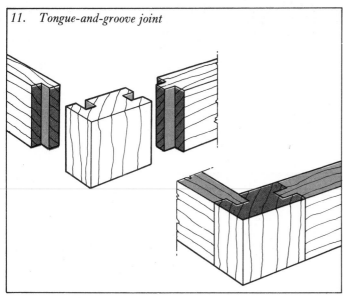

11. Tongue-and-groove joints are most often found along the edge of ready-made boarding. But a right-angled version of this joint is also found, for example, at the corners of boxes.

12. The end rabbet joint has a rabbet cut in one side to hide most of the end grain. This type of joint is quite often found in inexpensive cabinet work, because it is very easy to make with power tools.

13. The mortise-and-tenon joint is a very strong joint used to form T-shapes in frames. The mortise is the slot on the left; the tenon is the tongue on the right.

14. The through mortise-and-tenon joint is stronger than the simple type just described. It is sometimes locked with small hardwood wedges) driven in beside, or into, saw cuts in the tenon.

15. The haunched mortise-and-tenon is used at the top of a frame. The top of the tenon is cut away so that the mortises can be closed at both ends, and so retain its strength.

12. End-rabbett joint

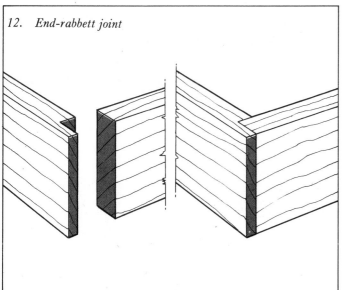

14. Through mortise-and-tenon joint

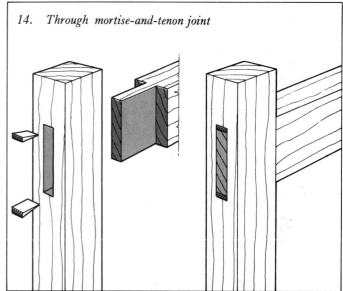

13. Mortise-and-tenon joint

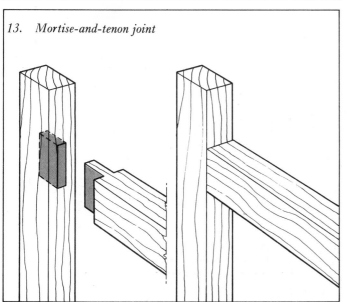

15. Haunched mortise-and-tenon joint

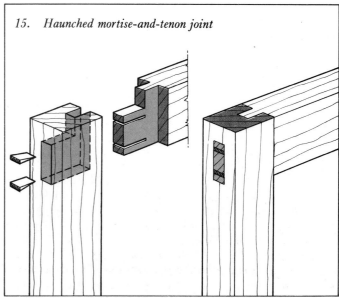

16. The bare-faced tenon is offset, with a "shoulder" on one side only. It is used for joining pieces of different thicknesses.

17. Twin tenons are used in very thick lumber. They give the joint extra rigidity and do not weaken the wood as much as usual. These are ideal for the construction of large items, such as sheds.

18. Forked tenons add rigidity to a deep narrow joint. The angled edge of the tenon is sometimes found in a haunched mortise-and-tenon joint.

19. Stub tenons are generally used on even deeper joints, but they are weaker and less rigid than the forked tenons just described.

20. The bridle joint is most often used where a long horizontal piece of lumber has to be fitted into the tops of several vertical pieces. This joint is best used for interior work as the end grain is exposed.

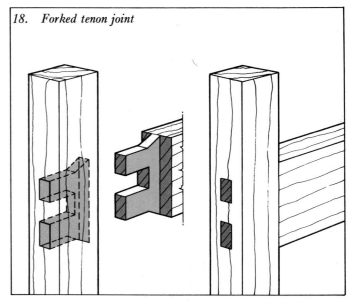

18.   Forked tenon joint

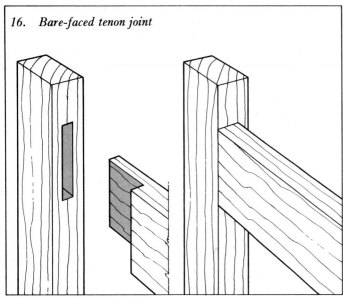

16.   Bare-faced tenon joint

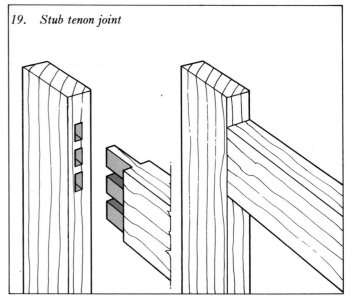

19.   Stub tenon joint

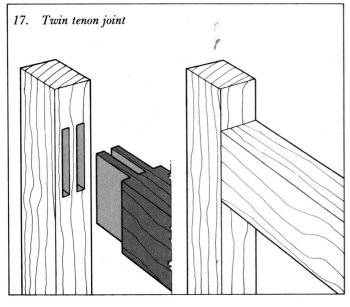

17.   Twin tenon joint

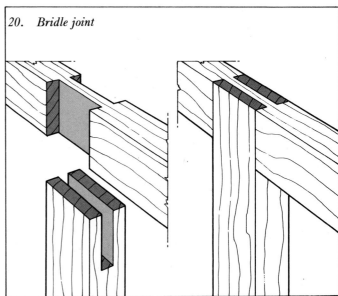

20.   Bridle joint

21. Box joint

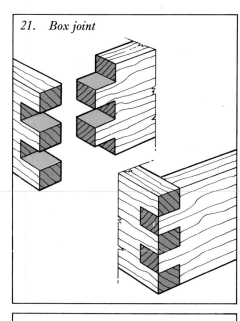

22. Single dovetail joint

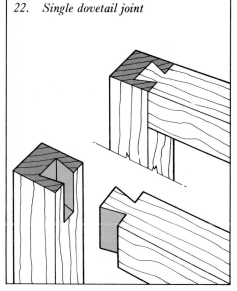

23. Through multiple dovetail joint

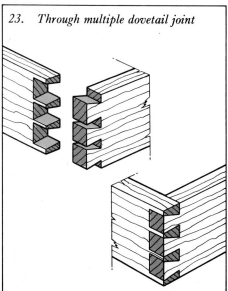

21. The box joint is quite strong and has a decorative appearance.

22. The single dovetail, like all dovetails, is extremely strong.

23. The through multiple dovetail is used at the corners of drawers where strength and good appearance are required.

24. The lapped dovetail is nearly as strong, and also has one plain face.

25. The mitered secret dovetail is also used in very high-quality work.

26. The stopped-lap dovetail is easier to make than a mitered secret dovetail.

27. The cogged joint is very strong and rigid.

28. The lapped scarf joint is used for joining frame members end-to-end.

24. Lapped dovetail joint

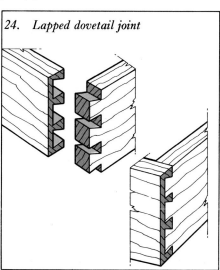

25. Mitered secret dovetail joint

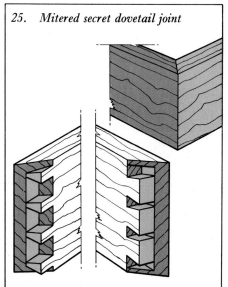

26. Stopped-lap dovetail joint

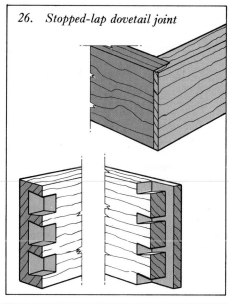

27. Cogged joint

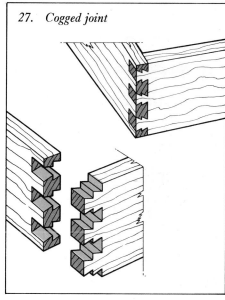

28. Lapped scarf joint

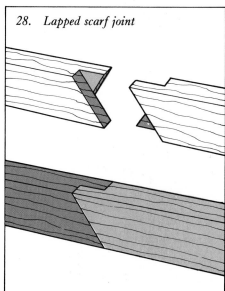

Workmate

For anyone who enjoys carpentry and wants to organize his workshop more efficiently, a Workmate® will prove a valuable addition to his equipment.

The Workmate® is not just a workbench, but a giant vise and sawhorse, as well as being a completely portable bench. It folds flat for easy storage and can, in fact, be hung on hooks on the wall. Its other great advantage is that instead of having to take the work to your bench, you can take the Workmate® to the location where you want to do the work, be it workshop, garage, any room in the house or even outside in the garden. Yet, despite its portability, any model of the Workmate® is strong enough to take loads up to 350lbs (159kg).

The basic construction is a framework, either of aluminum or pressed steel, with two long, flat vise jaws which enable it to be used either as a working surface or a vise. Independently operated handles at each end allow it to hold wedge-shaped objects – a big advantage over a con-

ventional vise. A series of holes is drilled into each of the vise jaws to take swivel pegs, which not only extends the maximum width of the object which the Workmate® will hold securely, but enables it to grip irregularly shaped objects firmly, even something as awkwardly shaped as a guitar! Horizontal V-grooves along the inner edges of the vise jaws permit tubular objects to be securely gripped. The more expensive models have a dual-height facility: a set of folding legs, which, when folded away, give the lower height which is more appropriate to use as a sawhorse and, when erected, raise the height of the whole structure to a more comfortable height for bench use. Also, on more expensive models, adjustable feet are provided at both heights to permit the Workmate® to be levelled on uneven surfaces.

Apart from its versatility in being capable of holding large objects – such as a door – or irregular shapes, another great advantage is that, unlike many con-

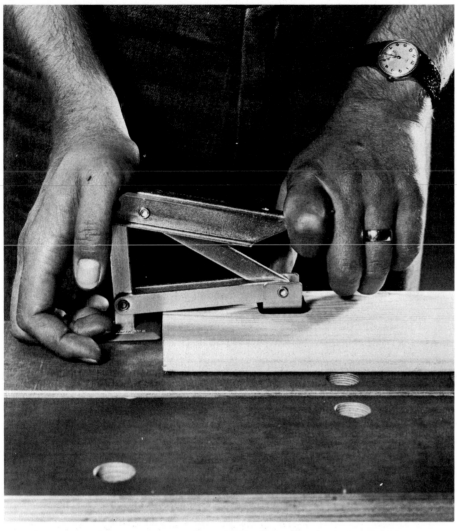

The Gripmate ™ clamp fits into holes in Workmate vise jaws to hold work pieces securely.

ventional fixed benches, you can walk around it to work on your project from any angle.

Attachments

To add to the versatility of the Workmate®, a range of attachments is available.

The Gripmate™ is a device for clamping work down to the working surface of the bench, similar in function to a C clamp. It consists of a vertical shaft, which, seated in a plastic liner, can be located in any of the swivel peg holes. A horizontal arm is lowered to the necessary level to grip the workpiece securely to the bench top, and a lever action then clamps it firmly in place. A rubber pad at the point of contact prevents damage to the work.

The Mitremate Saw Guide™ is designed for use with the Workmate® and any Black & Decker 7¼in (184.14mm) circular saw, and provides a much greater degree of accuracy than more traditional methods. It can be used for cutting mitres from 30° to 90°, and adjusts at increments of 5°.

When sawing at 30°, the maximum width of material is 7¾in (197mm), but this increases to 17in (432mm) at 90°. The base of the Mitremate™ clamps to one of the Workmate® vise jaws, and the protractor is set to the required angle. The workpiece is then held against the protractor, while the saw, clamped to the flat metal base, is fed into the work to give a clean cut and an accurate angle. The maximum thickness of board which can be cut is 2⅞in (73mm).

The Routermate Shaping Guide™ is designed to be used with most routers with a base-plate diameter of 6in (152mm) or less, to perform routing and shaping in conjunction with the Workmate®. The guide fence is clamped between the Workmate® vise jaws and the guide plate, with the router secured to it, slides along the channel in the guide fence. This can only be used with wood less than 1⅝in (41.3 mm) thick, but enables one to route edges, make beadings, and carry out your own freehand designs.

Power Tools

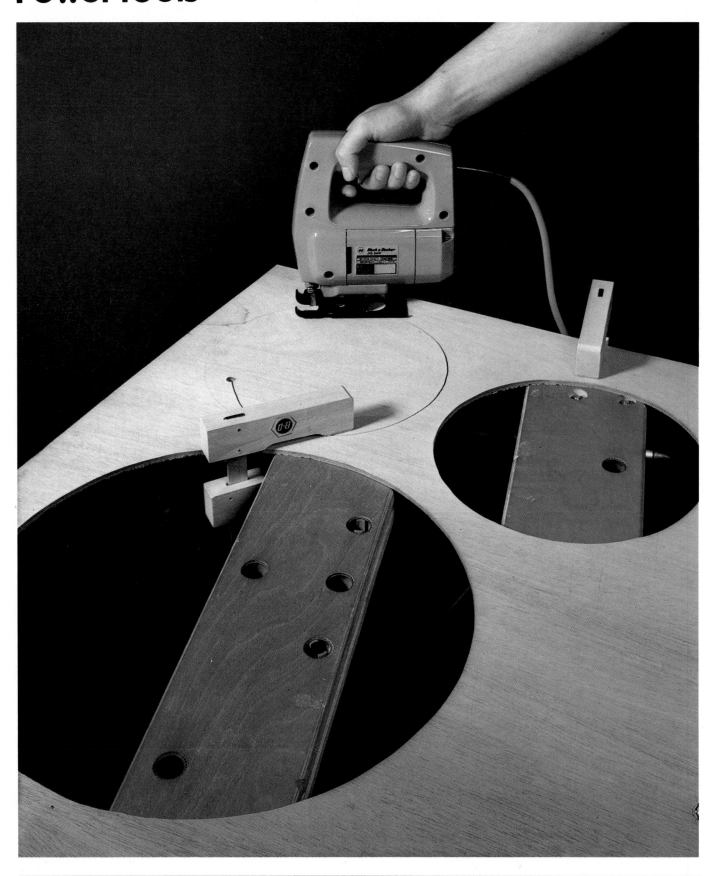

DO

KNOW YOUR POWER TOOL—read instruction leaflets carefully.

KEEP THE WORK AREA CLEAN—untidy work surfaces and benches invite accidents.

WEAR PROPER CLOTHING—do not wear loose clothing, aprons or unbuttoned cuffs, or jewelery which could get caught in moving parts. Rubber footwear is recommended for outdoor work.

WEAR SAFETY GLASSES—with most power tools.

USE THE SAFETY GUARD PROVIDED—keep it in place, and in good working order.

SECURE WORK—clamp down work, or use a vise. It leaves both hands free to operate the tool, and ensures that work cannot be snatched from your grasp.

KEEP TOOLS PROPERLY MAINTAINED—Keep blades and drill bits sharp at all times for optimum safety and performance. Ensure that tools are regularly cleaned and serviced.

DISCONNECT TOOLS WHEN NOT IN USE—so that they cannot be switched on accidentally. Check this especially when fitting attachments and making adjustments.

REMOVE ADJUSTING KEYS AND WRENCHES—make a habit of checking that all such adjusting tools are removed before switching on your power tools.

STORE POWER TOOLS SECURELY—when not in use, store your power tool in a dry, lockable place—away from children.

DO NOT

USE YOUR POWER TOOL IN DANGEROUS ENVIRONMENTS—in wet, damp or combustible atmospheres. Keep the work area well lit.

LET CHILDREN STAND TOO CLOSE—all onlookers should be kept a safe distance from the work area.

FORCE YOUR POWER TOOL—it will perform better and more safely at the rate for which it was designed.

ABUSE CABLES—never carry power tools by the cable, or tug it to disconnect the plug from the socket. Keep the cable clear of sharp edges, oil and heat.

OVER-REACH—keep a good balance and proper footing at all times.

START THE TOOL ACCIDENTALLY—never carry a plugged-in tool with your finger on the switch.

The drill

The Drill

Power drills can be used for far more than just boring holes. A large number of accessories can be fitted to the basic drill unit, enabling it to do many types of work that would otherwise require a specialized power tool.

Basically, a power drill is a compact electric motor fitted with a projecting shaft at one end on which is mounted a chuck, a revolving clamp that grips and drives drill bits or other attachments. The motor unit is held in the hand by a pistol grip and the motor is started by pressing a trigger at the top of the grip. For safety reasons, the motor stops if pressure is released on the trigger. But most drills have a locking pin that can be engaged to hold the trigger in the 'on' position.

Electric power is supplied to the drill by a cord that enters the machine through the bottom of the handle. On many modern drills, a complex system of insulation, known as double insulation, is built in to keep the user from getting an electric shock. In addition to extra user-safety, it is not necessary to earth such tools.

The motor is cooled by a built-in fan that draws air through slots in the sides of the drill. These slots must be kept uncovered and free of sawdust or the motor may overheat and burn out.

Should any power tool become hot, through heavy or prolonged use, the quickest way to cool the motor is to hold it safely away from yourself and the work, and run it at full speed in free air. This allows for the fan to provide maximum ventilation.

Many drills can be adjusted to run at different speeds. The normal type is a drill with a two-speed geared reduction, to run at up to 1,000 rpm and 2,500-3,000 rpm. These two speeds are suitable for most household jobs, and a machine with a two-speed gearbox is the best buy for the ambitious amateur. Some drills have been made with two speeds achieved electronically, through a diode switch. This is certainly a low-cost method, but the speed range is narrower, typically 1,700 rpm and 2,500-3,000 rpm, and there is some power loss at the lower speed.

Variable-speed drills, where the speed can be infinitely varied by an electronic device, are also available. This control can either be built into the trigger, working by finger pressure, or through a feedback system, an electronic chip

which enables a speed to be 'dialed' and then maintains it constantly, whatever the applied load.

Although with no practical use in carpentry, when choosing a drill, consideration should be given to drills incorporating rotary hammer action. These are most useful when drilling into brick, concrete or masonry is required. At the flick of a selector switch, the rotary hammer action can be engaged for easy drilling into such materials. As the drill bit rotates, it also hammers up and down, to break up hard aggregate in its path. Disengage the hammer selector for normal rotary drilling and attachment driving. This feature adds to the cost of the drill, but is well worth it for the additional scope it offers around the home.

Power drills come in various sizes, which are graded by the capacity of their chucks, that is by the largest drill bit that can be fitted into the chuck. Common sizes are $\frac{1}{4}$in (6mm), $\frac{3}{8}$in (10mm) and

$\frac{1}{2}$in (13mm). These refer to their drilling capacities in mild steel. In most cases, using a narrow-shanked bit, they will drill holes at least double these diameters in wood, and even more, using that bit. The larger machines have more powerful motors. A medium-sized machine should be adequate for all ordinary jobs. The $\frac{1}{4}$in (6mm) and $\frac{3}{8}$in (10mm) sizes fall into this category and are the most suitable for the household carpenter.

An indispensable accessory that every drill user will need is an extension cord. This enables him to use the tool at a distance from a power source. Cords are available in standard lengths from 25ft (8m) to 100ft (33m) or more, or you can make up your own. The longer the cable, the thicker the wires must be to prevent power loss. Larger drills also need heavier cords. You should have a ground fault interrupter if using a grounding type power tool outdoors. You can buy a portable GFI from large electrical supply outlets.

Drill bits and fittings

Many types of drill bits are sold for cutting different size holes in different materials. The most common sort are twist bits, used for drilling holes in metal and in wood up to the drill's capacity. The smallest common size of twist bit is $\frac{1}{16}$ in (1.6mm) and sizes increase in steps of $\frac{1}{4}$ in (6mm) up from this.

Larger holes in wood are drilled with spade type or Power Bore bits. The spade bit has a flat tip with a center point and two cutting edges. The Power Bore bit has a round cutting tip for a somewhat smoother cut. Forstner bits are used in a drill press. They cut neat, flat-bottomed holes, but they have to be withdrawn and cleaned out during the drilling.

Very large holes are drilled with spade bits (up to $1\frac{1}{4}$ in [31mm] or hole saws up to 3in [75mm]). The hole saw (also called a trepanning bit) has a revolving toothed ring attached to a central twist bit. The ring removes wood like a revolving pastry cutter. Different sizes of ring are available.

Very long holes, such as those up the shaft of standard lamps, are drilled with a power bit extension. This is a shaft held in the chuck with the bit held in a small setscrew chuck at the other end.

Other types of bit include countersink bits, for countersinking screw holes, and combination bits like the 'Screw Mate,' which are specially shaped to drill and countersink (or counterbore) a hole for a particular size of screw.

A plug cutter is often used in conjunction with screw work to conceal screw heads in wood. A bit the size of the screw head is used to counterbore a screw hole (to recess the screw head some way into the wood) and then a plug, like a short cylindrical piece of wood, is cut from a matching piece of wood, glued into the recess over the screw head, and planed flat to give almost an invisible result. The grain of the plug runs *across* its diameter.

For drilling hard masonry, special masonry bits are available. They look like twist bits, but have cutting tips made of a special hard alloy. If using a drill with rotary hammer action, then masonry percussion bits are required. These have specially hardened shoulders at the tip, to withstand vibration shock.

Glass and tiles are drilled with a spear point drill, which also has a hardened tip.

Drill bits for power drills have rounded shanks to fit into the chuck. For this reason hand-tool bits (for a bit brace) can't be used in a power drill. These

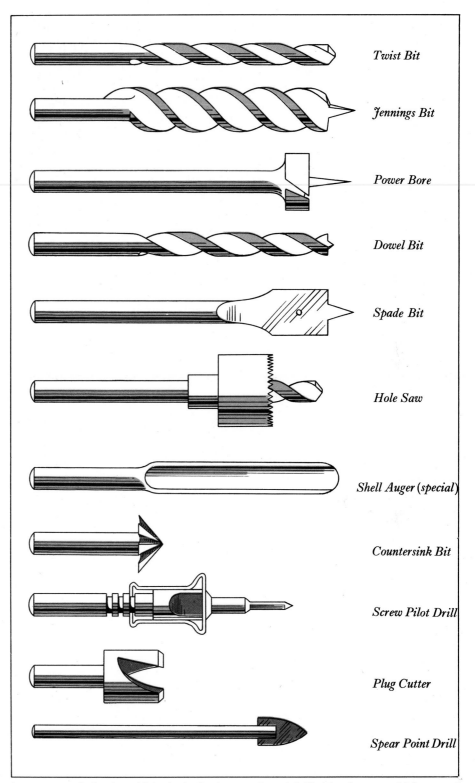

Twist Bit

Jennings Bit

Power Bore

Dowel Bit

Spade Bit

Hole Saw

Shell Auger (special)

Countersink Bit

Screw Pilot Drill

Plug Cutter

Spear Point Drill

have squared shanks which don't fit the power drill chuck.

Twist bits possess heads no wider than the shanks and are made in two types of steel, carbon and high-speed. High-speed steel is best for general use. It has a much longer life than carbon steel which dulls quickly when drilling metal.

Some drill bits have a **thickness** of plastic wrapped around the shank of the drill to provide depth indication. This can be moved up and down. You can make a homemade depth stop by sticking adhesive tape around the bit. Bright fluorescent tape is best. Drill to the lower edge of the tape.

Bit sharpener

Above: The electric drill bit sharpener has a special grinding wheel designed to sharpen drill bits at the correct angle. The bit drops into the appropriate hole in the top and is rotated gently while operating the machine. Tungsten carbide-tipped bits should not be used.

The bit sharpener

Sharp and accurate drill bits will produce better quality work, without undue time losses or frustration—and the bits lose their sharpness with continued use. You can restore the edge on a grinder or use special drill-bit sharpeners now on the market.

You can change the position of the bit inside the sharpener by an adjusting knob. In this way you can control the cutting of the first two angles. When using the sharpener the following procedure should be followed: set the adjusting knob on the front in the vertical position and insert the drill into the smallest hole of the top plate into which it will fit. If you insert it into a larger hole it will result in the bit being misshaped. The point of the bit should be in the center, otherwise the bit will drill oversized holes which will result in badly finished work and possibly waste of a valuable piece of lumber.

You can control the point position by the pressure you apply to the bit during sharpening. Turn the knob in a clockwise direction to increase the angle of cutting and in a counterclockwise direction to decrease the cutting relief.

Drill bits should be sharpened only when they are dry. When sharpening do not use water or any other liquid as a coolant. Sharpening makes the bit hot so do not touch it with your fingertips immediately after you have extracted it from the machine.

Only high-speed steel or carbon steel bits can be sharpened in the machine. Masonry, percussion and other bits with specially hard tips, or of a different tip-shape, should not be put into the machine, or they will be ruined.

Drill stands

There are two types of drill stand or press: the vertical and horizontal. With the vertical you can drill accurately at an angle of 90°. The horizontal stand is used mainly for driving rotary sanders, abrasive discs, polishing buffs and bonnets, grinding wheels and wire brushes. By adding a sanding table to the horizontal stand you can shape, sand and grind objects.

The vertical drill stand consists of a base with slots for bolting to a work bench or Workmate, a collar and clamping bolt for the upright column and the drill carriage. The carriage is raised or lowered by a feed lever and you can adjust for depth of hole and the travel of the drill on the column by adjusting the height of the collar which clamps on the column. A spring returns the carriage to the raised position.

When using the drill on the stand it is important to keep the cable out of the way by leading it away from the cradle of the drill. Lightly oil the column fairly frequently to prevent rust and also to enable the carriage to slide more freely.

Drill stand vises

This unit enables you to do accurate work on the base of the stand. Clamp the lumber to be drilled in the vise and then move the vise so that the center line of the hole to be drilled is exactly underneath the center point of the drill bit.

Tighten the vise by securing the bolts with a small wrench. Make sure it is tightly clamped.

Right-angle attachment

A right-angle attachment converts the drill so that you can drill at angles or in awkward places. To use it with the drill you may need adaptors. You can halve or double the speed of the drill (with some types) depending on which way around the attachment is fitted into the spindle of the drill. You may want to buy an extra chuck for the attachment.

You can also fit a grinding attachment to the chuck. But these should not be used at high speeds. Carefully follow the instructions of the manufacturer.

The right angle attachment enables you to drill at an angle of 90° to the drive line. The letters 'S' for slow and 'F' for fast are stamped on the flat face of the hexagonal drive adaptors at the ends of the accessory. When the drill is connected to S and the chuck fitted to F, the chuck rotates at twice the rated drill speed. When these two positions are reversed the speed is halved. Fast speeds are used for drilling small holes and for high-speed sanding.

The attachment can swing through 360° merely by rotating it by hand. The shape of the unit also allows access to drilling structural members such as floor joists, where the overall length of the drill bit and the length of the drill would exceed the space between joists. Place the tip of the drill bit at the point to be drilled and align the bit as nearly perpendicular to the work surface as possible.

Left: The use of a drill stand vise makes sure that the work does not move when using a vertical drill stand.

Bottom left: A vertical drill stand makes for greater precision, particularly where a series of holes is required.

Bottom right: A horizontal drill stand increases the versatility of your drill by holding it firmly when using such accessories as grinding or buffing wheels and polishing bonnets.

Drilling techniques

The great imperative when drilling is always to be careful that you are drilling at right angles to the work surface. The best and surest way of doing this is by using a drill stand. If you haven't got a drill stand you can use a try square, sighting the bit against the square.

A surer alternative is to use one of the jigs now on the market. Dowelling jigs, for example, give accurate siting and drilling of holes to take wooden dowels. These range from a simple drilled metal block to more calibrated and complicated types.

To position dowel holes without a jig, mark out the center of the hole on one of the two pieces of lumber to be joined. Insert brads into the marks. Nip the heads of the brads until only about $\frac{1}{4}$in (6mm) of shank is protruding. Then position and press the second piece of lumber down onto the first. Separate the pieces and remove the pins. You can then drill the dowel holes through the brad marks. Always make sure that the lumber is firmly clamped when you are drilling.

Large pieces can be held firmly by

Fig. 5

hand or even by foot. But always hold smaller pieces in a clamp or vise onto the bench or Workmate. Otherwise the drill may suddenly snatch the piece out of your hand and spin it around.

With a drill stand you can considerably widen the scope of your woodwork operations. Mortise cutting, for example is easy. After marking out the mortise you can drill out most of the waste to a finely controlled depth, finishing the resulting slot with a chisel.

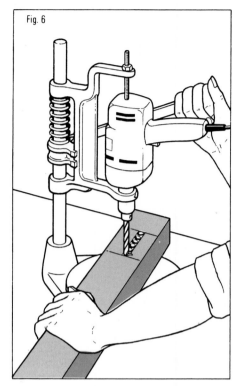
Fig. 6

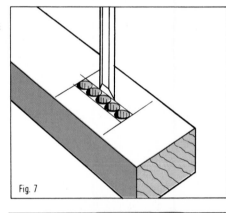
Fig. 7

Fig. 1

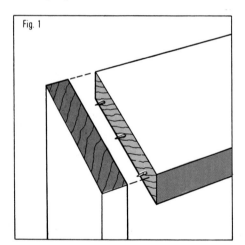
Fig. 3

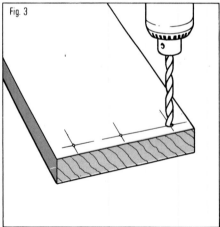

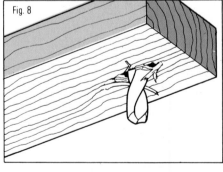

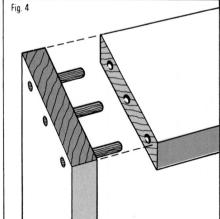
Fig. 8

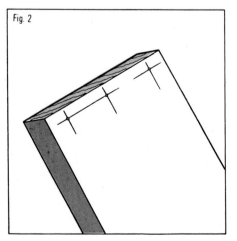
Fig. 2

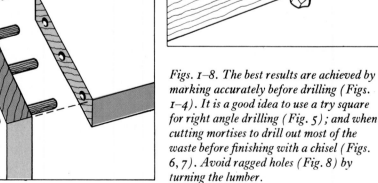
Fig. 4

Figs. 1–8. The best results are achieved by marking accurately before drilling (Figs. 1–4). It is a good idea to use a try square for right angle drilling (Fig. 5); and when cutting mortises to drill out most of the waste before finishing with a chisel (Figs. 6, 7). Avoid ragged holes (Fig. 8) by turning the lumber.

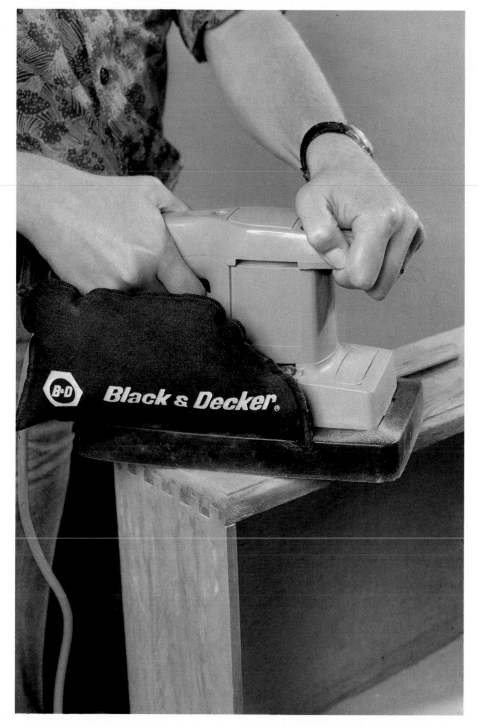

An orbital sander with dust extraction. This will produce a really smooth finish suitable for sanding, varnishing or even polishing.

Sanders

Several types of sander can be fitted to a power drill. The most commonly used is the disc sander. A flexible rubber disc is mounted in the chuck of the machine and an abrasive paper disc is fastened to it with a recessed central screw.

The sander is used at an angle so that only one side of the disc touches the surface being sanded. If the disc is laid flat against the surface or pressed too hard against the surface, it produces circular marks called swirl marks, which may be deep and difficult to remove. Even with the disc used at the correct angle, slight swirl marks are unavoidable.

A special type of disc called the 'Swirl-away' reduces these marks to a minimum. The disc is made of metal and is flat and completely rigid. To give it flexibility in use, the shaft on which it is mounted can be tilted (by means of a ball joint) at a slight angle while it is turning. This allows for normal power drill operation.

The drum sander consists of a wide revolving drum made of rubber, with an abrasive belt fastened around its perimeter. It makes no swirl marks, but should only be used for sanding curved edges of small objects or narrow strips of wood. On large, flat surfaces it tends to give an uneven result. This type of sander is also operated by a power drill.

The orbital sander, on the other hand, can be used to give a fine finish to any flat surface. It is an integral tool with its

Sanders

Right: Orbital sanders are available in ½- and ⅓-sheet sizes. This particular model can be fitted with a dust extractor. These power tools are suitable for use on lumber, plaster, most metal surfaces and paintwork. To achieve a fine finish, work through grades of sandpaper.

own motor and a large, flat sanding pad covered by an abrasive sheet. This moves backward and forward in a small circle and leaves no visible swirl marks. Many can also be adjusted for reciprocal (back and forth) motion.

The abrasive discs, belts and sheets for all these tools are available in many coarse, medium and fine grades as well as special types, such as 'wet-or-dry' and 'preparation' for rubbing down paint-work.

The finish on sandpaper depends on the number of particles or granules. The grains are widely spaced on 'open coat' paper to allow the dust to fall from the sander when it is removed from the lumber. This minimizes clogging when sanding paint or pitchy wood. The backing paper is tough and specially made to withstand the harsh action of the power sander. Note: ordinary sandpaper is not suitable for use with a power sander. Check with your local dealer.

Disc attachments

The disc attachment is the most widely used for rough sanding wood. It consists of a 5in (125mm) rubber backing disc which fits into the chuck of the drill.

Work with light sweeping strokes, starting with a coarse paper and working through all the grades to a fine paper. It is most important to keep the disc moving over the surface.

Drum sanders

Drum sanders are available in a wide range of widths and sizes. They can be used on convex and small flat surfaces, across or with the grain.

The arbor of the drum is fitted into the chuck of the drill. A tubular cover of abrasive material makes a band that fits around the cylinder. When the cylinder spins it is moved against the surface of the lumber. The abrasive is held on the drum by tightening a nut on the shaft, which expands the rubber.

Belt sanders

It is of great advantage to have an integral belt sander. The belt sander is powered by a motor housed in the sander body. There is an 'on' and 'off' switch and a second handle so that you can control it with two hands.

The sanding belt travels over two rollers, one at the front of the sander, the other at the rear, driven by the motor-operated roller, usually the rear one. To remove material fast, the sander can be used at about 45° to the grain of the wood, first pointed to one side, then the other. Keep the sander moving over the area to be sanded, so as not to make hollows in the surface. On rough work start with a fairly coarse sanding belt, and progress through medium to fine. Belt changing in most types takes less than a minute.

Power Sawing

Three basic types of circular saws are generally available, two of which are stationary shop tools. The most familiar form is the 'bench table saw'. It consists of a metal table with a circular saw blade protruding upward through a slot near the table center. The blade can be raised or lowered by means of a saw projection hand wheel. This makes it possible to set the blade height to the depth of cut desired. For example, you can set the blade to cut only half an inch into a thicker piece of wood, if you merely want to cut a slot in it, rather than cut all the way through it.

The blade can also be tilted to an angle to cut a bevel rather than a square edge. The angle of tilt is controlled by a saw tilt hand wheel. This angle of tilt in some saws, chiefly older ones, is controlled by tilting the table rather than the blade. Saws that control the angle of cut by tilting the blade are called 'tilt arbor' saws, those that control the angle by tilting the table, 'tilt table' saws.

The fence

Both types of bench, or table, saws mentioned are usually equipped with a 'fence'. This is an adjustable metal guide running parallel to the saw blade. It can be moved and locked to provide the desired space between itself and the blade, thus automatically setting the width of the piece cut, as when ripping a wide board lengthwise to make two narrower boards. In most cases the space between the blade and the fence should be measured with a ruler for accuracy, even if the saw has its own measuring scale. In all cutting of this type, however, allow for the thickness of the blade's cut (kerf), as it must be subtracted from the total width of the remaining wood.

The miter gauge

In addition to the fence, most table saws are equipped with a miter gauge. This is a guide that slides in slots parallel to the saw blade, and can be adjusted to any required angle. Thus, it is possible to set the edge of a piece of wood against the

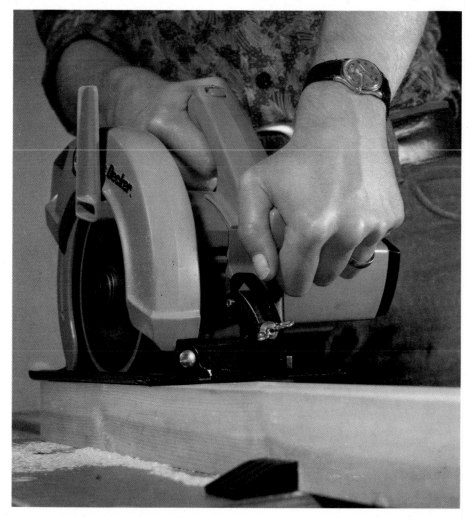

Left: A circular saw being used to rip a length of softwood. Always make sure that the blades are sharp; otherwise, these machines are liable to stall.

adjusted miter gauge, and push the piece through the saw so as to cut off the end of the piece at the pre-set angle. You can, for example, set the miter gauge at 45° to cut a miter, or at any other angle that happens to be required. Most miter gauges are also provided with removeable 'rod stops' that make it possible to cut any desired number of duplicate parts. As designs vary with the brand, follow the manufacturer's instructions in adjusting and operating the saw and its accessories.

The radial arm saw

The radial arm saw (often called a radial saw) does much the same type of work as the table saw, and is extremely versatile. Basically, it consists of a table at the usual height, with a vertical column at the rear, topped with an extending 'over arm' that supports the saw and its motor *above* the work. The over-arm assembly is made in several types, but their general operation is similar. Unlike the conventional table saw, the blade cuts *downward* into the work. For operations like miter cutting, the saw is swiveled to the required angle on the over-arm assembly, and locked, then pushed across the work on the over-arm's sliding track to make the cut. To make a square end cut, the blade is swiveled and locked at 90° to the front of the table, and pushed into the work, as described for miter cutting. For ripping, the blade is set and locked parallel to the front of the table, with the required distance between it and the fence, which is fixed at the rear of the table. Then the work is pushed through the blade. The blade can also be tilted to cut bevels. Many other operations, such as rabbeting and dadoing can be done on both the table saw and the radial saw. Because of design differences in different brands, however, these operations and others should be done according to the manufacturer's instructions. If you plan to buy either type of saw, it's best to examine it at the source from which you'll be buying, and look over the instructions at the same time.

The portable circular saw

With the portable circular saw you can do many of the jobs possible with the table saw and radial arm saw. But instead of moving the work through the saw, you move the saw along the work. The saw can be adjusted to tilt the blade to various angles to cut bevels. Another adjustment

raises or lowers the blade to cut to various depths, as in grooving work. On most portable saws there's also an adjustable rip guide that can be set and locked at varied distances from the blade to act like the fence of a stationary saw in ripping boards to specified widths. Angle cuts, as in mitering, may be made by guiding the saw along a marked line. Various types of miter and angle guides are also available. Before buying one, however, make certain that it can be used with the brand of saw you own.

Blade sizes

Power saw sizes are commonly designated according to the diameter of their blades, and the diameter varies over a considerable range, depending on the type of use. Typically, table saws likely to be used in the home shop take blade diameters from $7\frac{1}{4}$in (181mm) to as large as 12in (300mm). Radial saws are likely to be in the 8in (200mm) to 10in (250mm) blade diameter range. The portable circular saw ranges in blade size from as small as $4\frac{1}{2}$in (106mm) (for special work) to more than 10in (250mm). The smallest size, used for such things as trim, plastic laminates, and composition boards, may be available only on order from average outlets, as is often the case with the very large sizes, used for heavy work. A very popular size, with $7\frac{1}{4}$in (181mm) blade diameter is readily available at most tool and hardware suppliers. The important point in selecting any circular saw by blade size is its ability to cut the thickest material you're likely to use. A typical $7\frac{1}{4}$in (181mm) saw, for example, can cut through a piece of lumber 2 7/16in (62mm) thick at 90°, and through a $1\frac{7}{8}$in (46mm) piece at 45°. This is adequate to handle most house framing work and, of course, work on thinner materials.

Blade types

You buy blades for your power saws according to the type of work to be done. One of the most popular blades for general use in rough cutting is the combination or all-purpose wood cutting blade. This cuts both across the grain and with the grain. For plywood, you'll do best with a blade made for the purpose, with fine teeth that don't fray the surface. If you're cutting into wood that may contain nails you use a flooring blade. This type has specially hardened teeth that can retain their sharpness even if the blade shears

through occasional nails. Use this type of blade also if you're cutting used lumber that may contain nails. For light gauge metal there are metal cutting blades, and for heavier metals, there are abrasive discs that replace the blade. Similar discs are available for cutting masonry block and other like materials. If you need a blade for any special job or material tell your hardware dealer what the blade must do. He can then recommend the right blade for the job.

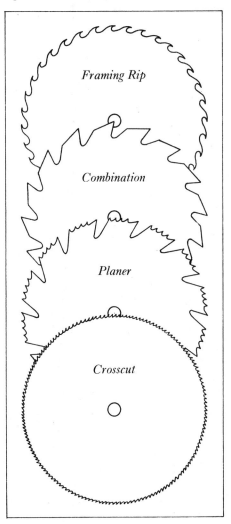

Framing Rip

Combination

Planer

Crosscut

About saw guards

When using any power saw equipped with a blade guard, use the guard on all work where its use is possible. On a typical bench, or table, saw, the guard (which covers the exposed portion of the blade above the table) can be used on practically all work where the blade cuts all the way through the wood, as in ripping and end cutting. In most cases where the cut is not a through-cut, as in grooving, however, the guard must be removed. The guard on a portable circular saw consists of two sections, the upper one fixed in position, the lower one retractable. As you push the saw into the work, as in cutting the end off a board, the retractable (spring-loaded) section of the guard springs upward as it comes in contact with the wood being cut, and snaps back in position as it passes over the far side of the wood. It can be used in practically all work except where the saw moves into the wood at a very fine angle. If this results in jamming the guard against the wood during entry, the lower guard may be retracted manually by means of a retracting lever on the guard. Follow the manufacturer's instructions on this, as in all phases of operation. And in all types of power saw operation, keep your hands well clear of the blade at all times, and never attempt repairs or adjustments with the saw plugged into the power outlet.

The portable jigsaw (also called a sabre saw)

This type of power saw cuts by means of a rapidly reciprocating straight blade that projects downward through the saw's base plate or 'shoe'. (This is the flat metal plate that rests on the work.) It can make either straight or curved cuts, starting from the edge of the lumber or from a hole bored through it away from the edge, as in cutting an opening in the mid-area of a plywood panel. It can also make its own starting hole. To do this, the saw is tipped, nose down, resting on the projecting front portion of the base plate, with the tip of the blade resting *lightly* on the surface of the wood. Then, after lifting the blade tip just out of contact with the wood, the saw motor is started with the trigger switch. When the reciprocating blade reaches full speed, the blade tip is *very gradually* brought into contact with the wood by tipping the saw backward. By tipping the saw farther backward, the blade is brought more deeply into the wood until it cuts all the way through. This is an operation that has many applications, but must be done with considerable care in order not to break the blade.

Blades

A wide variety of blades are available for the sabre saw, including fine-toothed types for cuts requiring a smooth finish, and coarse-toothed types for work where cutting speed is more important than smoothness. Both fine and coarse toothed metal-cutting blades are also available.

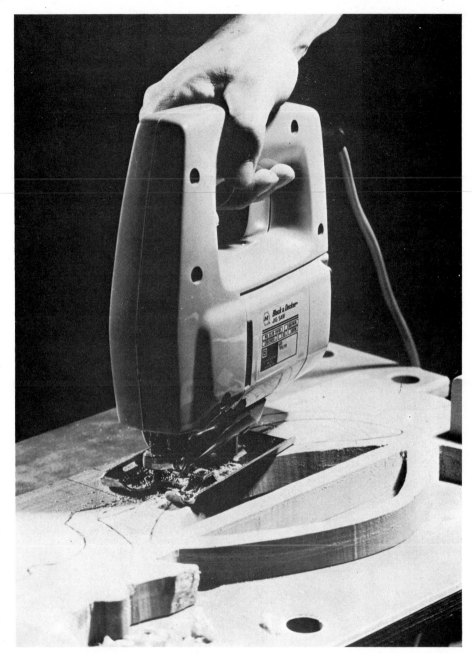

Above: A jigsaw used for intricate shapes.

The choice of coarseness in metal cutting is keyed to the thickness of the metal. At least two teeth should be in contact with the metal being cut, so that the work metal can't project between teeth, breaking them. In buying blades for a sabre saw be sure to select them to fit the saw you own. Some types are simply inserted into a slot in the lower end of the reciprocating shaft and clamped there by a setscrew. Others are held by a screw that passes through a hole in the upper end of the blade. In many cases, the package containing the blade specifies the type of use for which the blade is suited. If not, inquire where you buy. Choose narrow blades for small radius curves, fine teeth for smooth cuts, coarse teeth for fast cuts. Blades are also available with teeth on both edges for work where cutting in both forward and backward directions may be required. It's wise to keep a supply of the blades you use most, so that you have spares in the event of breakages. If you find your saw cutting more slowly than usual, unplug it and feel the tips of the teeth. If they're dull from wear, replace the blade. Using dull blades prolongs the saw's operation unnecessarily and results in more than usual wear of the working parts plus possible overheating.

Guides for portable saws

Ready-made guides are available for portable jigsaws (sabre saws) and for portable circular saws, though you may have to make inquiries to find one, as the type you want may be made by one of the smaller manufacturers. The more elaborate types enable a portable circular saw to accurately duplicate many of the operations of the radial arm saw. Typically, this type consists of adjustable tracks, along which the saw moves after the angle, width, or depth of cut has been set. Simpler forms may merely guide the saw in angle or circle cutting.

Some of these guides, such as the Trik-Trak, may be used not only with the jig saw or circular saw, but also with the router. Before buying, obtain full details of the guide to make sure the tools you

Right : Using a push stick for safety when working with small pieces of wood on a saw table.

plan to use with it, will fit. In the event you can't find a suitable ready-made guide, keep in mind the battens (like 1in x 2in [25mm x 50mm] stock) clamped to the work can also serve as accurate guides. A clamped batten, for example, lets you make a long straight cut (that might otherwise be wavy) with a sabre saw.

Sawing techniques, bench saw

It is dangerous to feed small pieces of wood into the blade with your hands, because your fingers get uncomfortably close to the blade and the slightest slip may cause a serious accident. Wood is very likely to slip on a bench saw because of possible flaws in the wood, and the tremendous torque of the blade. This is a problem especially when using the miter

gauge at an angle.

Small pieces of lumber should be pushed towards the blade with a push stick, a piece of lumber with a V-shape cut out of the end so that it can hold the piece of wood firmly. It does not matter if the push stick gets cut because you can make another one in seconds. Fingers are not so easily replaced.

Mitering

Mitering, or cutting wood at a 45° angle to make a mitered joint, can be done quickly and accurately. To cut a miter across the face of a piece of wood, as if making a picture frame, set the protractor on the miter gauge accurately to 45°. Then lay the wood against the guide and slide the guide and wood together town the table into the saw blade.

As a general rule, when crosscutting the lumber should be held with both hands on one side of the blade and the offcut allowed to fall away freely. If you push from both sides, the pressure tends to close up the cut around the blade, causing it to jam and buck dangerously. If you must hold both sides, apply pressure in such a way as to hold the cut open.

To cut a bevel or miter along the edge of a piece of lumber set the tilt protractor to a 45° angle. On some bench saws the table tilts instead of the blade. Slide the wood along the fence towards the blade in the normal way, but be sure the cut as it is finally produced is of the correct width you require.

Taper ripping is cutting a very shallow taper on a long length of wood so that it is narrower at one end than the other. It is used, for example, when cutting legs for tables, so that the finished result is smooth and graceful.

Taper work is best done by making an adjustable jig out of two moderately long battens. Set them face to face and fasten them together by a hinge at one end and a slotted metal strip fastened with wing nuts on the other. By moving the free ends a distance apart and locking them at this distance with the strip and wing nuts, the jig can be set at any shallow angle.

In use, the jig is slid along the fence together with the wood to be cut. This method is particularly convenient when there are a large number of identical designs or shapes that must be cut, to make sure they match exactly.

Left: The miter guide enables accurate cuts across lumber at any angle.

Housed joints and dadoes

Housed joints and other grooves can be cut simply and accurately by setting the saw blade to the required depth of cut and cutting the sides of the groove first, using the fence to keep them straight. Then slide the fence away and remove the wood between the cuts by repeatedly passing it over the saw blade. Mark the extent of the groove on top of the wood or you may cut past the edges. Except with very narrow grooves or housings, this method is faster than chiseling the whole thing by hand, although for a stopped dado you will have to cut the last inch or two by hand since the curved saw blade cannot reach the inside corner of the joint. It is also more accurate because the depth of cut is constant all over the groove. When cutting tenons, wood can be removed in the same way. (A stopped dado is one in which the groove does not extend all the way to the edge of the wood.)

Rabbets

There are two ways of cutting rabbets on a bench saw. One way is to cut along one side of the rabbet, using the fence to ensure accuracy, and then turn the wood through 90° and cut the other side.

This involves two operations for each rabbet. A faster way is to mount the blade on 'wobble washers', a pair of angled washers that make the blade wobble from side to side as it revolves. As a result, the blade cuts a wide groove instead of a neat line. The width of the groove is restricted by the size of the slot in the saw table, because if the blade 'wobbled' too far it would cut the table. But you can always make several passes to cut a wide rabbet.

When you have set the blade on its washers, fasten a piece of scrap wood to the fence to protect it and move it until the blade just brushes the scrap wood at the apex of its wobble. Now any piece of wood that is slid along the scrap wood will have a rabbet cut out of it the same width as the wobble of the blade—or even narrower if you adjust the blade to cut farther into the temporary scrap wood fence. The depth of cut can still be adjusted in the normal way.

Great care should be taken when using wobble washers because the oscillation of the blade makes it more difficult to see in operation. At all costs, keep your fingers well away from it.

Kerfing

Kerfing is a special technique that enables a piece of solid wood to be bent in a curve. Rows of close parallel cuts are made across the wood on the inside of the curve through half to three-quarters of the wood's thickness, and all the way along

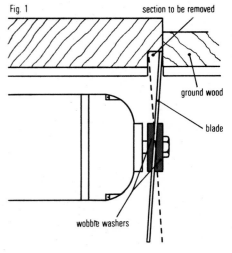

Fig. 1. A circular saw fitted with a "wobble washer" enables you to cut slots in lumber.

Figs. 2–4. Cutting a housed joint using a circular saw. Fig. 2 shows the first cut being made on the far right-hand side of the section to be removed. A further series of cuts is made on the left-hand extremity. The remaining stock can be removed with a chisel.

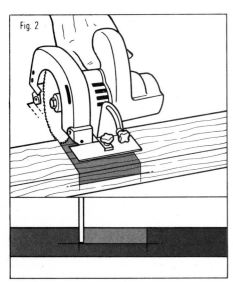

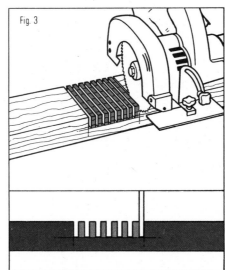

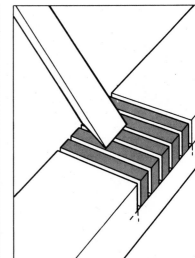

the part that is to be curved. The wood can then be bent and it helps if you wet or steam it as well. Use a crosscut or planer blade to make the cuts. A combination blade is too coarse and may break out the wood between cuts.

Kerfing reduces the strength of wood sharply, and should not be used for load-bearing frames. It is really only suitable for outside curves, with the saw cuts on the narrower radius. The wood could be bent the other way but the surface would probably wrinkle unattractively. When used correctly kerfing produces a neat curve that is more difficult to make by other methods.

Tenon joints

To cut a tenon on a bench saw, first cross-cut the lumber to length and make sure that the ends are square. There are two ways of making a tenon. In some circumstances the first is less satisfactory.

The first method uses a series of cross-cuts. First adjust the depth of cut and use a batten as a guide. Make the first cut on the shoulder of the tenon. Then make a series of crosscuts along the length of the marked tenon and complete. Lap joints can also be made in this way.

The second method is the more obvious. Make a crosscut and an endwise cut for each shoulder. When making the endwise cut use a push stick with your right hand to slide the tenon towards the blade.

The rip fence will need two settings so that you will save time by completing all the crosscuts before completing the endwise cuts. The corresponding mortise can be made by the method described in the chapter on the power drill.

Left: The use of a saw table for long cuts leaves both hands free to guide the lumber.

Left: Setting the depth of cut on a circular saw.

Fig. 1. A tenon being cut using a bench-mounted circular saw.

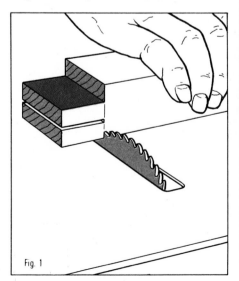

Fig. 1

The jigsaw (sabre saw)

A power driven jigsaw is used for the same jobs as a coping saw, that is for cutting curves and complex shapes. Its blade is small and pointed and moves rapidly up and down with a reciprocating motion. Various types of blade are available for cutting wood, plastic and sheet metal. But they will not cut very thick boards or sheets. Typically they can manage a 2in (50mm) thick softwood board or hardwood half as thick.

Jigsaws should not be pressed forward too hard or the highly tempered blade may snap. But they should be held firmly down onto the material you are cutting to resist the downstroke of the blade.

The average blade is narrow enough to cut $\frac{1}{2}$in (13mm) radius curves but will not turn a right-angled corner. It can, however, be started in the middle of a piece of wood by tilting the machine forward on its nose and gradually lowering the blade into the wood until it is upright.

Jigsaws are available in a variety of sizes and power ranges. Tables are available for some, making it possible to use them as stationary tools, with the blade pointing upward through the table.

Always make sure that the blade is mounted tightly and that it remains that way. If you should drop the jigsaw, if it is hand-held, never attempt to catch it. This may sound obvious but your instinctive reaction will be to grab.

Cutting

The jigsaw has a built-in blower to prevent wood dust from obscuring the marked line you are sawing along. Air is directed from the drill to behind the saw blade through a tube.

Always support large panels of wood to prevent the usual pinching and jamming of the blade. If there is any vibration stop and check the wood supports before proceeding. To start a cut, rest the base plate on the surface of the wood and slowly move the blade toward the edge of the lumber.

General operation

In using the sabre saw always hold the saw firmly on the work, which should be solidly supported. Be sure that the blade can cut along the required course without striking the work support. (If the work is resting on saw horses you don't want to cut the saw horse.) If you're cutting into the work from the edge, keep the saw

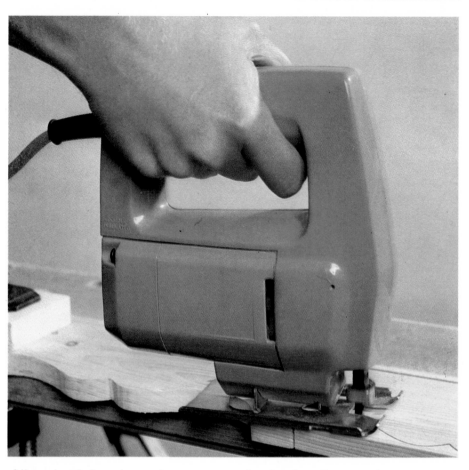

A jigsaw is at its best when cutting curves and intricate shapes.

blade a short distance from the edge (with the saw resting on the work) when you start the saw. Allow a few seconds for the saw to reach full speed before you move the blade into the work, and move it gently. Don't force the saw along the cut; move it at a rate that results in a steady cutting speed without noticeable slowing of the motor. Low powered saws will, of course, cut at a slower rate than high powered ones. Suit the rate of cut to the saw. Practice on scrap material before you tackle an actual job, if you've never used a sabre saw before. Usually, it takes only a few minutes to get the feel of the saw and the knack of using it. In making curved cuts, keep the saw moving forward and steer it along the line to be cut. Don't try to make right-angle turns. If you must cut out an opening, as in plywood, with sharply squared corners, first cut it out with rounded corners *inside* the square-cornered out-line. Then use the saw to square the corners by cutting inward to the sharp corner from both adjoining sides. In all sabre saw work, keep your hands out of the saw's path, and do not

place hands or fingers under the work, as the blade projects down through it.

General cuts

A rounded cut can be cut in one careful operation but several passes are needed for oblongs and squares. The first side should be cut to its fullest extent before bringing the blade back down the cut and curving it gently away from the cut to carve out the second side. The piece left in the corner can be cut out later. Keep the motor running throughout this procedure. Cut the remaining sides in the same way.

A key-hole size opening can be cut by moving the blade backwards and forwards making slight stabs at the wood.

The saw can also be used to cut straight lines although a circular saw is preferable. Keep long lines accurate by using a clamped batten as a guide.

Pocket cutting

To start a cut in the middle of the lumber, tilt the saw forward and allow it to make its own starting hole, as described earlier. This works less well on thicker wood and it may be necessary to drill a hole first before inserting the blade.

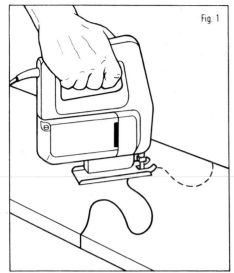

Fig. 1

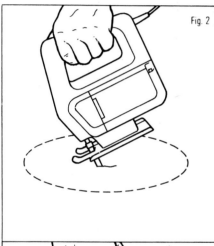

Fig. 2

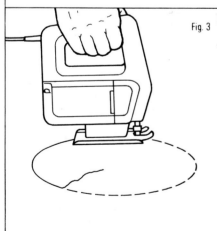

Fig. 3

Keep the saw vertical when cutting a curved shape (Fig. 1). Make a pocket cut into the center. Cut to the edge, then around the perimeter (Figs. 2, 3). A jigsaw is very useful for general sawing and trimming jobs (Fig. 4), particularly where a section has to be removed from the material of a surrounding surface area (Fig. 5).

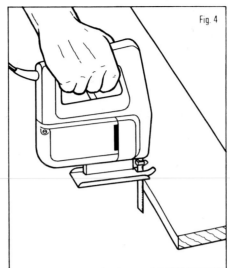

Fig. 4

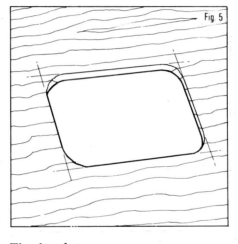

Fig. 5

The bandsaw

The bandsaw is a machine designed for cutting curved shapes primarily in wood; however, special blades can be obtained for cutting plastics and metals. It can also be used for straight cutting. A number of bandsaws are available either as bench or floor models. They all have the same basic components. The type and size selected will largely depend upon individual requirements.

The bandsaw consists of a continuous length of narrow flexible blade which is mounted like a belt running over bandwheels, one of which is driven by an electric motor. The bandwheels have tired peripheries—these tires are an integral part of the wheel and are slightly convex. This convexity allows the blade to run centrally on the wheel and counters the tendency for the blade to run off. In order to keep the blade running vertically through the saw table and to support the blade when cutting, upper and lower guide assemblies are provided—the upper

being vertically adjustable to allow for varying thicknesses of lumber. Both guides should be adjusted so that they will almost touch the blade, a piece of notepaper between the blade and guide will provide sufficient clearance. The upper guide assembly will also have a thrust roller placed at the rear of the blade to prevent the saw blades being pushed away from the work. The adjustment should allow the blade to run clear of the roller when not cutting, again a thin sheet of cardboard can be used as a feeler to check this. The blade must run under tension and to enable tensioning to be carried out, the upper bandwheel is adjusted vertically by a tensioning knob on top of the bandsaw. To enable the saw to run accurately on the crown of the bandwheels, a side knob, which is locked after setting by a lock nut, can be used to set the verticality of the bandwheel. Correct tension can be ascertained by checking the blade at a point between the upper and lower guards, i.e. just above the table—it should have a maximum of flex. The experienced user will instinctively be able to tell that the blade is in correct tension. Typically, the blade should have a flex of about $\frac{1}{4}$in (6mm) in a 6in (150 mm) span.

The saw table may be fitted with a slot in which a miter guide can slide. This guide is used when cutting miters and angles and also when cutting off battens. Most bandsaw tables can be tilted for angled cutting.

Before commencing to cut a particular job it is as well to make several cuts in a waste piece to ensure even and accurate cutting. Mark a straight line on a piece of lumber, set the upper guide to just allow the passage of the lumber. Switch on the machine and cut along the line; no great pushing effort should be required. If the saw doesn't cut accurately, incorrect setting of the guides, incorrect tension or bad sharpening and setting of the blade may be the cause. Crosscutting is easier than cutting with the grain; always begin a curved cut at a point on the cross-grain. Never backtrack, if the blade should get to the position where no further cutting can take place, stop the machine and withdraw the lumber.

When ordering a band, remember to state the length of the band, its width and the material you intend to use it on. The following points should be borne in mind when deciding the saw blade width for a typical home shop bandsaw, such as a

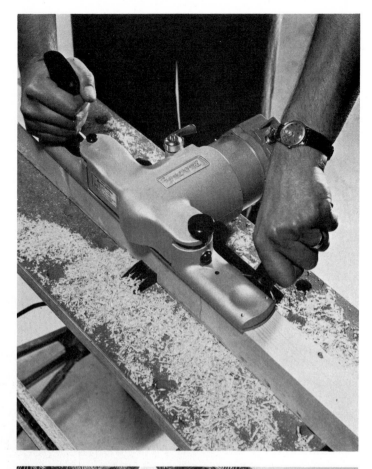

Right above: A powered plane is used in the same way as a conventional plane, but with much faster results.

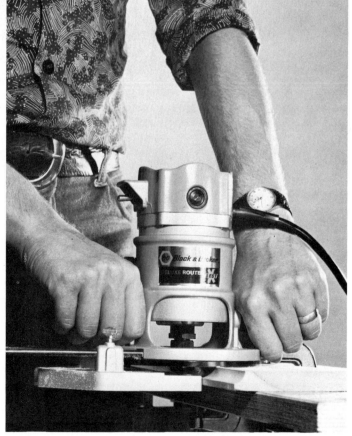

Right below: A router can be used with various cutters to apply shape and finishing detail to woodworking projects.

14in (350mm) model. (Size based diameter of the band wheels—sometimes called pulleys.) Basically, the smaller the radius to be cut the narrower the blade required. For $\frac{1}{4}$in (6mm) radius you'll need a $\frac{1}{8}$in (3mm) blade; for $\frac{3}{4}$in (18mm) radius you can use a $\frac{1}{4}$in (6mm) blade, a popular size. For additional sizes of radii, you will have to check the manufacturer's specifications.

The planing machine (jointer)

One of the most useful of the workshop machines, it can be used not only for planing wood straight, flat and smooth, but also for rabbeting, tenoning and chamfering. The planing machine is usually referred to as the jointer, because of its use in jointing (straightening) wood surfaces. It can also be used to make some types of molding and to plane wood to a specified thickness.

Cutters may be honed in the machine by resting an oilstone on the rear table, suitably protected by thin cardboard and rubbing lightly across each blade. The blade must be held in its highest position and the cutter block locked with a small wooden wedge. When badly blunt and possibly nicked they will have to be removed and reground. Grinding is a job best left to the expert.

Using the jointer

Follow the same rules when using the jointer as when planing on the bench. Examine the lumber for loose knots and always plane with the grain. Generally take thin cuts, particularly where the lumber appears to be coarse and cross-grained. Feed the lumber carefully, at a steady rate of feed, particularly when planing end grain. Should the lumber be planed on all four edges always plane the end grain first, any slight tearing at the corners will be removed when planing the long grain. When planing end grain only, begin cutting from one edge then reverse the wood to complete the cut from the the other edge.

When planing, wherever possible use a push stick, and keep the cutter guard in position leaving sufficient clearance to push the wood underneath. To plane boards wider than the planer, several cuts will need to be made. When planing a curved edge remove the 'hump' first. When cutting a short taper, use a support stick for the lumber and pull the work over the cutters. To cut a long taper use a push stick on the lumber.

The router

The router is undoubtedly one of the most versatile machines available to the home worker, but unfortunately much of its versatility remains unexplored.

Grooving, rabbeting, molding of all kinds, curved cutting, including disc cutting, housing, dovetailing, tongueing, mortise and tenoning, carving and a host of other processes are possible.

The basic machine consists of a motor, complete with a collet chuck which can be fitted with a very wide range of cutters made either of high speed steel or tungsten carbide tipped for cutting man-made boards. The machine fits into a body or housing to which is fitted a means of adjusting the depth of cut and also a method of locking the setting once it has been made. A straight and circular fence can be fitted to the body, and to the fence a device can also be attached for cutting circles. The fence is drilled to allow a wooden fence to be secured either for greater depth or to lengthen the fence for greater security when cutting. A dovetailing attachment together with special dovetail cutters permits the joining of wide boards in cabinet construction.

Using the router

The method of securing the cutters and depth adjustment varies according to the type of machine; the carpenter would therefore be advised to read the manufacturer's instructions with great care before attempting to use.

Unlike any of the other machines discussed the speed of the routers is extremely high, often up to 26,000 rpm. At these speeds, with cutters in good condition, the quality of finish both with and across the rain is extremely good. When using molding cutters it should be noted that some of these cutters have pins or pilots placed centrally below the cutter. The pilot runs on the edge of the lumber and thus limits the cut to the exact contour of the router bit. Care must be taken in use. Should the machine be transversed too quickly, the motor will be overloaded, conversely too slow a rate of feed will result in the lumber being burned and possible drawing of the temper of the tool. If cutting to the full thickness of the lumber, when using pilot-type cutters, it will be necessary to attach an additional piece of wood underneath the piece being cut, to provide a running edge for the pilot. The machine must always be moved from left to right when

cutting, with the lumber held securely in a vise. Always wait for the machine cutter to come to rest before returning the router to the bench.

The handyman who wishes to increase the versatility of the router may use an accessory 'shaper' table made for some routers, complete with an adjustable fence. This will prove to be most versatile when long lengths of lumber have to be cut. Great care must be taken to keep the hands well away from the router bit when using the machine in this way.

The bench grinder

One of the most useful tools in any workshop is the bench grinder. It enables all edge tools to be resharpened quickly—and sharp tools make for easy work. If you have a grinder attachment for a power drill, you will have to set it up every time you want to sharpen something—which is often. In practice, this means that you will not sharpen things often enough. So a bench grinder, which is not particularly expensive, is a good investment.

Another advantage of an integral bench grinder is that it has two revolving shafts—one on each end of the motor.

Chisel and plane blades are sharpened in two operations: grinding, to get the blade the right shape; and honing, to put an edge on it. Different grinding wheels are needed for each operation, so having both of them on the same machine speeds up work considerably.

Sharpening is done against the front curved edge of the grinding wheel, and not against the flat circular face. The wheel revolves so that the front edge moves downward. This keeps sparks and fragments of metal or abrasive from being thrown upward into the eyes (but *always* wear goggles). Adjustable tool rests are provided in front of each wheel to hold blades steady while they are sharpened.

Cutting wheels come in various grades. For most jobs, a medium wheel for grinding and a very fine one for honing should be all you need. Special extra-tough wheels are made for honing the hardened tips of masonry drills.

The wheels are fastened to their shafts by nuts screwed onto the threaded ends of the shafts. The wheel on the left has a left-hand thread to stop it from coming undone in use. The wheel on the right has a normal right-hand thread.

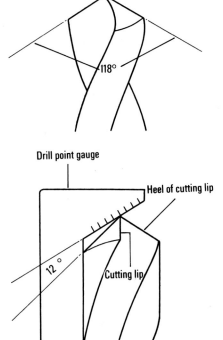

Above: Maintaining the correct angles is essential for accurate sharpening of drill bits.

Sharpening chisel and plane blades

Chisel and plane blades, though completely different in shape and use, are sharpened in exactly the same way. In both types of blade, the preliminary grinding to shape of the edge of the blade should give the ground surface an angle of 25° to the flat face of the blade. Then it should be honed at the slightly greater angle of 30°. The 5° difference saves you from having to hone the whole ground surface. Only the tip is honed.

To sharpen a blade, first lay it on the tool rest of the grinder with the point touching the stationary wheel, and measure the angle where the point touches. Move the blade until the angle is 25°, and memorize the position of the blade. Now take the blade away, start the wheels and lay the blade lightly against the coarse wheel. High speed and light pressure are the secret of good grinding. Move a wide blade from side to side across the wheel, so that its whole edge is ground evenly.

Grind on one side only until the blade is properly shaped, when the length of the ground surface should be 2½ times the thickness of the blade. Every few seconds of grinding, remove the blade from the wheel and dip it in cold water to stop it from overheating. An overheated blade 'loses its temper' and turns blue. If this happens, grind off the blue part.

The freshly ground surface will be slightly hollow in shape because of the curve of the wheel, but that doesn't matter. The next stage is to hone it.

Find the correct angle of the blade against the stationary wheel as you did before, except that it should be 30° and not 25°. Then start the grinder and lay the sloping side of the blade against the fine wheel—but only for a few seconds. The wheel will turn the edge of the blade over, producing a fine 'burr' on the other side. Cool the blade and lay the flat side *flat* on a flat stone and slide it back and forth to turn the burr the other way. Then turn the chisel around again then give the other side a few strokes at 30°. This will turn the burr again.

Continue doing each side alternately, using very light pressure and reducing the honing period each time. Eventually the burr will break off, leaving a razor edge.

Blades can be honed several times before they lose their shape and have to be reground.

Sharpening twist drills

Twist drill bits can be sharpened either on a drill sharpener like the motorized Black & Decker 7980 bit sharpener or the 79-800 bit sharpener, that is o perated by a power drill, or on the workshop grinding wheel with the aid of a drill bit grinding attachment. The attachments, available from hardware stores, are made in several forms. Some are manufactured as complete units, including a small grinding wheel, to be powered by an electric drill. Others are designed to clamp on the tool rest of a motorized workshop grinder. Whatever the type, follow the instructions of the manufacturer, as procedures vary with the brand. It is also possible to sharpen drill bits freehand, as experienced machinists often do. But don't try it without the experience. The basic principles, however, apply to all methods of drill bit sharpening.

Before your first sharpening job, study the point of a new bit, preferably a fairly large one, so you can see the details. If you think of the pointed drill tip as a cone, the usual angle at its apex will be 118°—59° on each side. This is satisfactory for drilling most materials, including soft to medium steel. (You can buy a flat metal drill point gauge at large hardware stores to check this angle.) The angle, however, is measured at the cutting edges of the drill tip, often called the cutting lips. The drill must be ground so the metal behind the cutting lips slopes downward away from the lips so the drill can bite into the work. So the angle measured at the trailing edge behind the lip (also called the heel) of each cutting lip is approximately 12° less than at the tip. The principle is simple. The cutting lip must be in contact with the material being drilled, and no other portion of the drill tip must be rubbing on the material in such a way as to keep the cutting lip out of contact. Think of the two cutting lips at the drill tip as two chisels rotating in contact with the material being drilled.

For extensive drilling in specific materials angles other than 118° are sometimes used for peak efficiency. For wood and thermoplastics, an 82° angle may be used. For very hard steel it may be 135°. For general use, however, you'll avoid complications by using the 118° angle. As you gain experience you can try the variations mentioned if you work in the materials to which they apply.

Drill bits are ground against the flat, side surface of the grinding wheel, not on the perimeter, in order to produce straight cutting edges. Be sure the wheel is suitable for this type of use. The drill is rotated during the process so as to grind the tip all the way around, forming the conical shape. The clearance is then ground behind each cutting lip individually, by tipping the bit as necessary. Use the method recommended by the manufacturer of the drill grinding attachment, or the drill bit sharpener.

Some of the tools in this section will only interest the professional or the ambitious amateur who wants to indulge his hobby to the fullest. All of the tools will vastly increase the range of projects you can tackle, and the standard of work will be even higher.

The lathe

A lathe adds a whole new dimension to your carpentry skills. For one point, you move beyond plain, square shapes, and you can make round objects of any contour and with as much (or as little) decoration as you like.

Wood-turning lathes are available in a range of sizes, from small ones powered by electric drills to huge professional models that can make sections for newel posts that are several feet long. But if you can afford a reasonably large-sized lathe you will find it much more versatile and useful in the long run than a miniature one.

All lathes are designed in much the same way. The main frame, to which all other parts are attached, is called the *bed*. The size of the lathe bed controls the maximum length of the piece of wood that can be fitted into the lathe. On the smallest self contained lathes, it may be as short as 12 in (300mm) or less. A typical workshop lathe will have a much longer bed, however, for greater usefulness and practicality.

The *headstock* is a strong support

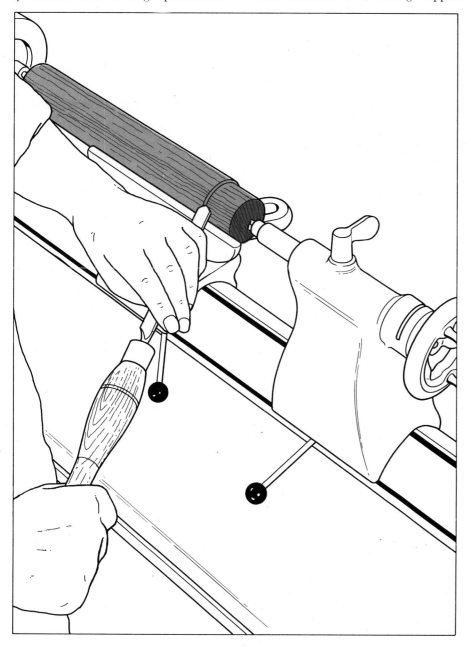

Left: A wide variety of cutting tools is available for use in wood turning. The angle at which the cutting tool is held is critical to the final result. Great accuracy is required when setting up the lathe.

mounted at the left-hand end of the lathe bed. It houses a revolving cylindrical spindle. This is threaded on the outside and has a hole down the middle so that fittings to hold the workpiece can be attached to it in two ways as described below.

At the other end of the lathe bed is another support, the *tailstock*, used for the turning of long objects such as lamp stands. The tailstock can be slid along the lathe bed to suit the length of the workpiece and can be fixed in any position. In *spindle work*, the workpiece is clamped between headstock and tailstock and spins on a *dead center*, a flat plate with a blunt central spike, fixed to the tailstock.

In *face plate* work—the turning of wide, flat objects such as bowls—the tailstock is not used and the object is fastened to the headstock only. This is why the headstock spindle provides two methods of attachment. Spindle work is fixed on a *morse tapered spur center* inserted into the hole in the spindle. Face plate work is fixed firmly with large wood screws to a flat face plate, and the face plate is in turn screwed onto the threaded end of the spindle.

On the near side of the lathe bed there is a *tool rest*. The tools used for shaping the wood are held against, and slid along, this rest, which can be adjusted and moved in all directions to suit any type of work.

Lathe tools

Special tools are made for lathework. There are three main kinds: gouges chisels and scrapers. At first sight, they may look like ordinary bench chisels, but there are important differences and bench tools should never be used for lathe work.

Lathe tools have very robust blades and extra-long handles which are usually made of hardwood and have strong brass ferrules.

All lathe tools are supplied ready-ground to shape, but you have to sharpen them yourself. This must be done in a special way for each type, as described below.

Gouges are used for roughing, or cutting wood roughly to shape. A gouge blade is U-shaped at an angle of 40° around the outside of the U. It can be ground in two shapes: straight across, which makes the gouge suitable for cutting flat, open surfaces, and with the

corners ground further back than the center, which makes the gouge right for cutting inside curves such as the insides of bowls.

Gouges come in sizes from $\frac{1}{4}$in (6mm) to 1in (25mm) wide. The wider sizes are used for roughing straight lengths and the narrower ones for sharp inside curves. In addition, the $\frac{1}{4}$in (6mm) gouge can be used for boring, such as is done in the middle of a bowl to mark the depth to which is should be cut (see below).

Chisels are used for finishing work once it has been roughed out with a gouge. They have a straight cutting edge, which may be ground square across or on the skew. The range of widths is the same as for gouges, but unlike gouges (and ordinary bench chisels) they are ground on both sides at an angle of 15°, so they come to an edge at 30°. Skew-edged chisels are more common, because they are easier to hold in the correct position. Both types, however, are used for the same purpose.

There is also a special type of chisel called a parting tool, which is made in one size only. The blade is like that of a chisel, but its end is ground in a V-shape rather like a spear. It is used for cutting a finished piece away from the waste wood left at its end. It is also useful for marking out a block before you start cutting it to shape.

Scraping tools are a form of chisel, but are used for fine finishing work. They are ground at a very shallow angle. One popular and useful type of scraper has a round 'nose', or sharp edge. Another commonly available type has a shallow V-shaped nose which most people regrind to suit their particular needs. Best bet: use the original shape if you are a beginner.

All lathe tools are normally sold with their handles. If you do happen to buy some in 'blade only' form, make sure that the handles you buy for them are the special extra-large lathe tool handles and not ordinary chisel handles. These ordinary chisel handles are too small either to accept the tang of a lathe tool blade or to hold firmly when working.

Grinding and sharpening

When you buy lathe tools, they are generally ground to shape and sharpened. The method of sharpening differs from that for ordinary bench gouges and chisels: the sharpening angle is the same as the grinding angle so that, when they are sharpened, an equal thickness of metal is removed from all over the ground edge, and its surface has a single bevel. Sharpening in this way takes longer, and needs more care, than the conventional method, but is absolutely necessary if the tools are to be used in the right way.

You will need several abrasive blocks for sharpening your chisels, gouges and scrapers: two oilstones with flat surfaces (coarse and fine) and a shaped oilstone (slipstone) of a suitable size to fit the inside curve of your gouges. One stone should be used for gouges, which wear stones unevenly and another kept exclusively for chisels and scrapers, which need a perfectly flat surface. Slipstones are available with an egg-shaped cross-section, so one will be suitable for all the curves on your various gouges if you choose one of the correct size.

Gouges should be sharpened at an angle of about 40°. To do this, hold the gouge with one hand at each end and apply the point bevel-side down to the stone with its bevelled edge exactly parallel to the surface of the stone—you can feel the angle by raising and lowering the handle until the bevel lies flat on the stone.

Run the gouge backwards and forwards over the stone and roll it from side to side at the same time, so that all of the curved surface of the bevel touches the stone on each pass. Since you are sharpening the blade only on one side, a burr will appear on the upper, visible side of the edge. If this burr appears all around the curve it will show that you are moving the blade in the right way.

As soon as the burr has appeared all around the edge stop sharpening and rest the blade, still bevel side down, on the tool rest of the lathe or some solid surface. Then lay the round slipstone flat in the groove of the gouge and slide it carefully over the edge to remove the burr—this should happen quite quickly.

Chisels should be sharpened on both sides at 15°. Apply the chisel to your perfectly flat oilstone with its bevel resting flat on the stone as before. Then move it backwards and forwards (but not, of course, rocking it) until a burr appears as before. Turn the chisel over and sharpen the other side in the same way. Remove the burr very carefully on the flat surface of the stone.

Scrapers are sharpened in the same way but on one side only and at the original angle. Remove the burr by laying the unground side of the scraper flat on the stone.

Preparing the wood

Wood that is to be turned is normally square in cross-section. It is time-consuming and messy—and can be dangerous—to cut the corner off entirely on the lathe, so it has to be trimmed very roughly to a circular cross-section before you begin.

A piece of wood that is to be spindle-turned should be about 2in (50mm) longer than the finished article to allow for the waste at either end to be cut off when turning is finished.

Find the exact center of the wood at each end by marking the diagonals from corner to corner. Then draw the largest circle on this center that will fit on to the end of the wood and plane the corners off the wood all along its length to make it roughly octagonal in cross-section. Take care not to cut below the line of the circle at any point.

Use a tenon saw or a chisel to prepare one end of the lumber to receive the spur center, making sure you get a tight fit perfectly centered.

Prepare the other end for the dead center of the tailstock by denting the center mark with a center punch, and apply a little oil or grease to the mark to make the wood revolve freely.

The wood is now ready for clamping between the headstock and tailstock. This is done simply by tapping the spur center well home onto its slots, screwing it on the headstock, sliding the tailstock firmly up to the other end and locking it in position. Adjust the tool rest as close as possible to the wood, and turn the wood around once by hand to make sure that it does not catch. The rest should be set just above the center line of the lathe.

Wood for face-plate turning should be just over 1in (25mm) thicker than the finished work. Plane one face perfectly flat, draw the diagonals and the circle as before, and saw off the corners of the block nearly down to the line of the circle. Lay the face plate exactly over the center of the block, mark and pre-drill screw holes in the block through the holes in the face plate, and fasten the plate firmly to the block with stout 1in (25mm) screws.

Screw the block and plate on to the spindle and set the tool rest as close as possible to the face of the work, but just below the center line.

Safety precautions

Never wear loose clothing when working on a lathe; anything that gets caught will be rapidly wound into the machinery, taking you with it. The ideal clothing is a buttoned-up overall, but it should not have any holes in it; these can be particularly dangerous. Remember to use goggles.

Just in case something does happen, make sure that the 'off' switch of your lathe is placed so that you can reach it in a hurry, without looking. Don't allow anybody to stand near you when turning wood.

Before you start work, always make sure that the workpiece is firmly fixed to the lathe, the lathe is firmly clamped down and all its parts are secure. These last two points are especially important.

Lathe speeds

The ideal speed for a particular turning job depends both on the operation being performed and on the diameter of the workpiece. The larger the diameter of a workpiece, the faster is its speed (in inches per second) at the outside edge for a given lathe speed (in revolutions per minute). So for most ordinary operations (other than end boring, finishing and final parting off) the smaller the piece the faster it should turn. A list of ideal speeds is given below.

If you are using a miniature lathe, you are most unlikely to be able to vary its speed to this extent. Many very small ones have only a single speed around 3000 rpm. This is all right for miniature turning, which the very small lathes are designed to do.

The solution to this problem is to buy a variable speed lathe for your shop, but there is an important point to observe in choosing one. Any speed-change pulley systems require space above, below, or behind the lathe. If you're short of space, select a self-contained lathe with built-in motor and electronic speed control—typically by means of a dial on the front of the headstock. This type requires no pulleys, belts, or auxiliary 'jack shafts'.

Another point to watch is that the speed control includes the range you'll require.

Turning techniques

When removing wood with a gouge or chisel, the tool is held point slightly upward on the tool rest with its bevel or grind tangent to the revolving surface.

The edge of the tool is then tipped slightly in the direction the tool is moving to make a clean shearing cut. This is why the edge has to be sharpened at the same angle at which it is ground. If there were a double slope the handle would be likely to catch and possibly fly out of your hand or damage the work. For this reason it must be held with both hands.

Note that the full width of the cutting edge is not used but only the section of the edge toward the direction of movement. The tool (in finishing) should be applied to the workpiece at one end (or side for end plate turning) and moved smoothly across to the other side (or the center) in a continuous movement. If you stop, it will create a ridge that will be difficult to remove.

Finding the correct angle and movement requires careful practice, so you should try the techniques out on worthless scrap lumber until you get them right.

In spindle turning, the length of the finished object should first be marked onto the prepared lumber with the parting chisel. This tool should be laid flat on the tool rest and brought into contact with the workpiece to mark it with a neat V-shaped groove.

After the lumber has been marked, it should be roughly cut to shape with a gouge; in some cases the parting chisel can be used. The gouge can be held flat on the rest and the center of the cutting edge used to cut, or tilted at an angle and slid across the work point first, depending on the contour of the part you are cutting. Be very careful that the far, upper edge of the cutting surface does not catch on the wood through incorrect angling.

Once the work is roughly cut to shape it should be smoothed. The correct tool for this is a skew chisel. Whether you use a straight or skew-edged chisel is entirely a matter of preference.

When you switch from gouge to chisel, adjust the tool rest so the chisel can make a clean cut on the workpiece and again move the rest to the wood as possible. You must stop the lathe to do this. Then restart the lathe and lay the chisel on the tool rest with the skew of its edge (if any) tilted towards the center line of the workpiece, and the edge itself angled at about 45° to the center line. Raise the handle until the end of the edge nearer the tool rest comes into contact with the wood, then move it across smoothly as before.

Final smoothing is done with a scraping type tool such as a 'spear', round nose, or square nose chisel, which are used in a different way from the other tools. The tool rest remains slightly below the center line and the scraper is laid on it horizontally and bevel-side down, so that it meets the work at a scraping rather than shearing angle. It should be used with great care and the overhang between the tool rest and the workpiece should be kept to an absolute minimum to keep the tool from vibrating.

Face plate turning is done in a slightly different way. The sides of the gouges used should be ground well back to make them fully round-nosed, or they will catch. You will be cutting into a vertical surface instead of one that slopes away from you, so the tools are positioned for scraping action.

A tool should be applied to the wood as usual, but it should not be angled to cut in, as you may be working on end grain. Instead, it should just be pushed gently towards the wood in scraping position, which will start the cut more smoothly.

When you are working across the face of a block, you will find that when you get to the middle, the rotation of the block tends to twist the tool around. Even if you hold it firmly to resist this, it will still not cut properly. There is a special technique to overcome this difficulty.

First use the parting tool to draw a circle $\frac{1}{8}$in (3mm) from the center of the block. Then mark the intended depth on the blade of a $\frac{1}{4}$in (6mm) round-nosed gouge by sticking on a piece of adhesive tape. Start a cut with the gouge about $\frac{1}{2}$in (13mm) from the center and cut towards the center in the conventional way. When the trailing edge, which is doing the cutting, reaches the marked circle, swing the handle of the gouge around smoothly so that it is at 90° to the face of the wood. You can then use the gouge as a drill to cut a hole into the wood until the tape is level with the surface.

diameter of work	nature of work	lathe speed (in rpm)
under $\frac{3}{4}$in (19mm)	roughing and general turning	2600-2800
$\frac{3}{4}$in (19mm) - 7in (180mm)	,,	1400
7in (180mm) - 12in (300mm)	,,	900
all sizes	sanding, burnishing	2600 - 2800
all sizes	end boring, lathe drilling, parting off	280 - 300
under 2in (50mm)	roughing	900-1300
	general turning	2400-2800
2in (50mm)— 4in (100mm)	roughing	600-1000
	general cutting	1800-2400
6in (150mm)— 8in (200mm)	roughing	400-600
	general cutting	800-1200
all sizes	sanding, burnishing	2600-2800
all sizes	end boring, lathe drilling, parting off	280-300

Spray painting

Spray painting is a technique which, if used correctly, can save you a lot of time and trouble in decorating a house or painting furniture. It is suitable for both inside and outside work, and is particularly useful for painting rough exterior surfaces such as stucco, which are hard to paint with a brush. On wood it gives a finished, even feel to the furniture you have made.

The type of paint you are planning to use should be thinned according to the manufacturer's instructions. Most conventional paints are suitable for spraying. Gloss paint sprays well but requires care to avoid getting sags and runs all over the surface. Masonry paints, as used for the outside of houses, can be sprayed only with special equipment including an extra-wide nozzle (to let the stone particles through) and an automatic stirrer (to stop them from settling). In some areas this equipment may be rented.

The state of the weather will affect painting outdoors. Obviously, it must be prevailingly dry. No exterior painting, by any method, can be done if it keeps raining. But spraying is also affected by wind. If it is a windy time of year in your part of the world, do not use a spray gun outside the house.

Apart from this difficulty, exterior spray painting is easier than interior spray work. There is less to mask, for one thing. Doors and windows must be masked, but drainpipes, for example, can be left unmasked and brush-painted, afterwards, on top of the sprayed coat, in the color of your choice. Lawns, flower beds and paving at the bottom of the wall can simply be covered with a tarpaulin or weighted down polythene sheet. Hard surfaces such as paving can be masked with a thin layer of earth, which is brushed off afterwards.

There are no special tricks about preparing lumber for spray painting. Exactly the same techniques are used as for any other kind of painting. Holes and cracks in the wood should be filled and, if necessary, the surface should be primed. The primer can be sprayed on if necessary, but mask the areas not to be painted and read the section on cleaning out the spray gun before you begin.

Masking and protection
Sprayed paint gets everywhere. Anything that you do not wish to be sprayed

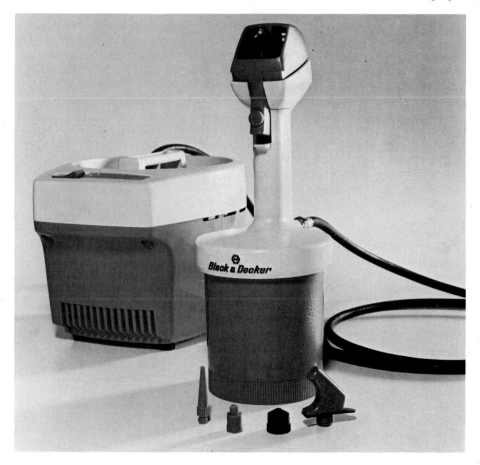

Left: Spray painting will give a far more professional finish than hand painting. A compressor provides the air power for the spray gun, which can be fitted with varying nozzles according to application.

Spray painting

The more thorough the preparation, the better will be the finished, sprayed article.

must be masked thoroughly before you begin—and 'thoroughly' is the operative word. It is not good just hanging a sheet of newspaper in front of an object, because the fine mist of paint will easily float around the back of the paper. It must be properly wrapped in newspaper and the edges of the paper stuck down all the way around with masking tape.

Proper masking takes quite a long time and uses a lot of paper and tape.

Fortunately, masking tape is not expensive—but do not try to save money by buying cheap tape. Inferior grades of tape stick too well and pull the paint off the things they are stuck to. Good masking tape will save you a lot of trouble. Buy two widths: ¾in (19mm) for holding down newspaper, and 2in (50mm) for covering small objects such as door handles and pipes.

Interior and exterior woodwork, drainpipes and other objects that are going be gloss-painted later in the course of redecorating do not need careful masking, and may need none at all. They will get some overspray but will be painted later. Natural stonework, on the other hand, should be masked with great care, because it is very hard to get paint off it. Indoors particular care should be taken to protect the floor, especially if there is wall-to-wall carpeting.

Furniture that cannot be taken out of the room should be stacked in the middle and well covered with drop cloths. A quick way is to use plastic sheets held to the floor with a stapler. The staples come out easily when the job is done. Make sure there are no gaps that paint mist can float through.

When painting indoors, take care to give the room proper ventilation. Mask windows in the open position. You should also protect your lungs by buying a painter's mask and plenty of replaceable pads for it, because the paint clogs pads up quickly. Do not laugh at this precaution; it is really necessary. Paint spray can be harmful to your respiratory system.

Wear old clothes for spray painting. This applies to garments that would not be touched by ordinary painting, such as socks. Remember to take off your wristwatch and any rings or bracelets you may be wearing.

The equipment

Whatever you are using it for, most spraying equipment should operate at a pressure of around 40lb/sq in (2.8 kg/sq cm). This does not apply to the kind of low-pressure spray gun that attaches to a vacuum cleaner.

Spraying techniques

All types of paint must be diluted with the appropriate thinner to make them suitable for spraying. Emulsion paint should be thinned with water; gloss paint with turpentine, or a substitute. Most of these are very flammable, SO DO NOT SMOKE when spraying with it, and turn off all pilot lights when spraying any type of flammable paint.

The exact amount of thinner to add to any type of paint can best be found by experience, but follow the manufacturer's instructions. Too little makes the paint too thick, so that the nozzle clogs in a few seconds. Too much makes the paint so thin that it does not cover the surface properly. Experiment on scrap material till you get it right. Even when the paint is the right consistency, the spray nozzle will probably clog occasionally. It should be cleaned out with a suitable solvent.

Spray the wall with horizontal strokes of the gun, holding it 12-18in (300-450mm) away from the surface. Move the gun back and forth parallel to the wall, rather than sweeping it in an arc. Spray on only a thin coat or the paint will run; you may have to put two or three coats on the surface, but spraying is so quick that you will not waste much time doing this. Here, again, follow the manufacturer's instructions.

Every time you stop spraying, even for a few minutes, clean the gun thoroughly with the appropriate solvent (the same as you use for diluting the paint). This is important—once the paint dries, you will have a much harder time.

Projects

Hall stand

If you have ever rummaged for your overcoat among a pile of coats hanging on a solitary hook in the hallway, or stumbled over carelessly left shoes, then this hall stand would be ideal for you. It is not only easy but also has plenty of space for all the family's coats as well as room to store shoes and umbrellas. The lower part of the stand can be used as a telephone table or a stand for a flowerpot.

All the major parts of the hall stand are cut from one standard 4ft × 8ft (1219mm × 2438mm) sheet of ½in (13mm) birch plywood. You will need a small extra sheet of plywood for piece F, which is ½in × 14½in × 30½in (13mm × 368mm × 775mm) (Fig. 3). The only other wood you need is 1in × 1in (25mm × 25mm) or 1in (25mm) triangular molding. The clothes rack of the hall stand can be made from 1in (25mm) dowel, but a length of round tubular steel will carry the weight of the coats better.

The construction of the hall stand, with its butted joints, is straightforward. The only difficult job is accurate butting of the pieces from the sheet of plywood. A power jigsaw is an essential tool for this aspect.

Marking the panels
Accurate marking of the plywood sheet is essential, or the finished construction will not be square. To mark the sheet you will need a tee square, a large framing square, and a hard pencil. You will also need a steel tape measure and a straightedge, or a 3ft (1m) rule.

First, make sure that the plywood sheet is perfectly square, by measuring the diagonals. If it is not, you will have to plane it square with a jack plane. Then you can mark out the sheet, taking all dimensions from the marking out diagram (Fig. 1). Mark out the board in a series of straight lines, marking the right angles from the edges of the sheet with the tee square. Check that the lines are parallel with the edges of the sheet. Note that ¼in (6mm) has been allowed for the width of the saw cuts.

All the curves shown in Fig. 1 have a radius of 4in (102mm). The best way of marking these is to make a cardboard template of the curve. Draw a circle with a 4in (102mm) radius on a piece of cardboard, and then square a horizontal and a vertical line through the center point. Cut out the circle.

At the points where the curve is to be drawn (Fig. 1), measure a distance of 4in (102mm) from the corner along the two lines that form the right angle. Lay the cardboard template on the sheet of plywood with the ends of the squared lines touching the marked points. Then mark around the template to form the curve in the corner.

Cutting the components
You will need a power jigsaw to cut out the panels for the hall stand. You must cut accurately, and follow the order of cutting shown in Fig. 2. To help make the straight cuts you can clamp a wooden straightedge to the plywood, along the marked lines. This will guide the blade of the jigsaw. When you come to the end of the first cuts, and start to make the second cuts, curve the saw out into the waste area of the plywood (Fig. 2). These corners can be trimmed square with a crosscut saw later.

When you have cut the panels for the two large sides of the hall stand, clean them up. Trim off any waste plywood. Then lay the two panels together, one on top of the other, and clamp them with a few C clamps. Plane the edges square with a block plane and shape the curves neatly with a spokeshave or comparable trimming tool. Then unclamp the boards and mark the inside faces. The shaped 'inside' edge of the two large sides should be rounded slightly – this looks better than square edges. Don't do this on any of the straight outer edges of these sides, though, as the other components of the hall stand are butt-jointed to these.

The other panels for the hall stand can now be cut from the plywood sheet. Refer to Figs. 1 and 3 for the dimensions. Trim the panels square, ensuring that the 15½in (394mm) dimension is exactly the same on all panels.

There is an alternative method of construction, if you are not sure that you can cut two identical panels for the large sides. This method uses two sheets of plywood which are clamped together and cut, after the shape of the large side has been drawn onto one of the sheets. This, however, wastes more wood than the procedure described above. Another

Hall stand

method, which involves altering the dimensions of the unit, is to cut one 4ft × 8ft (1219mm × 2438mm) plywood sheet in half across its width into two 4ft (1219 mm) panels and cut out the outline of the sides, with scaled down dimensions, with the two panels clamped together. This will give you a hall stand about 3ft 10in (1169mm) high – ideal for children's overcoats.

Assembling the unit

All the joints used in the construction are butted and nailed with extra reinforcing provided by 1in × 1in (25mm × 25mm) strips of lumber. If you use square-sectioned lumber for these reinforcing strips, you should cut it to a triangular cross-section. Alternatively, buy triangular-shaped molding. You will have to miter the ends of the wood at joints, so do not cut the pieces of reinforcing strip yet – 'direct measure' and cut them to length just before you nail them in place.

Lay one large side of the unit on a flat surface with the best face down and glue and nail the strip to it with 1in (25mm) brads in the position shown in Fig. 3. Miter the ends of the strip where necessary. The outer edge of the strip should be flush with the edge of the sides except at the back of the canopy where F fits. Here the molding must be set in ½in (13mm) from the edge so that piece F fits flush with the back edge of the unit.

When you do this job on the other large side, make sure you nail the strip to the inside surface to make the sides a 'pair'. Lay the second side down on the flat surface with its long edge butting that of the first side top to top. Then nail the molding in place on the second side.

The next step is to nail the other panels to the two large sides. This job can be a little awkward, because the large sides are unwieldy, so you may need some help. Nail piece A (Fig. 3) from the table, or lower, section in place, and then pieces E and F from the canopy, or upper, section. Piece C from the table section comes next, followed by pieces B and D, the tops of the table and canopy sections of the unit.

The umbrella stand

The next stage is to cut and assemble the umbrella stand. This is a simple box construction, glued inside the two large sides.

Cut the sides for the box out of scrap lumber 4in × 16in (102mm × 406mm).

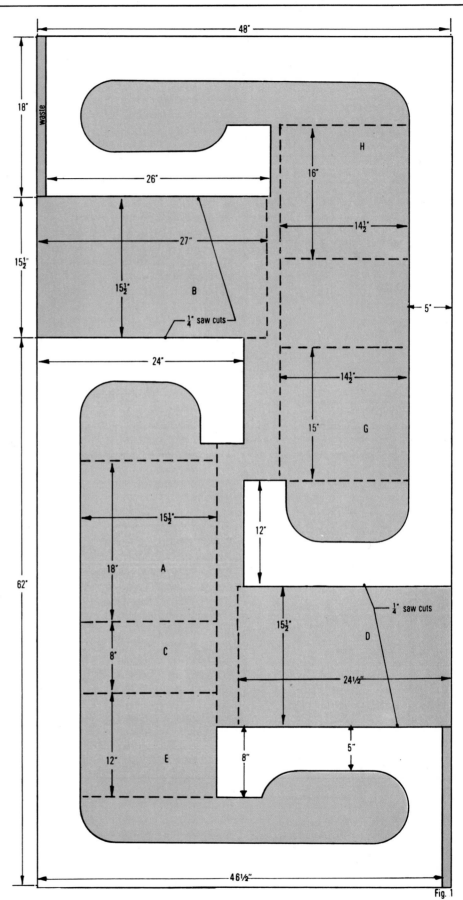

Fig 1. The method of laying out the panel for the hall stand.

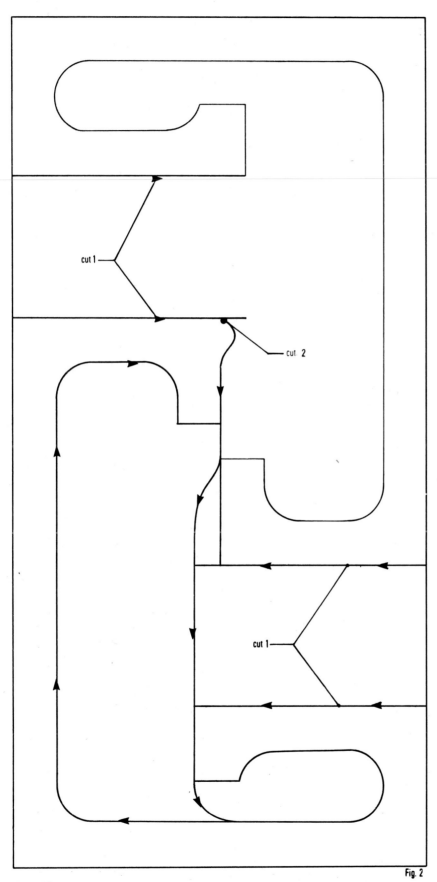

Fig 2. *The sequence of cutting the panel for the hall stand.*

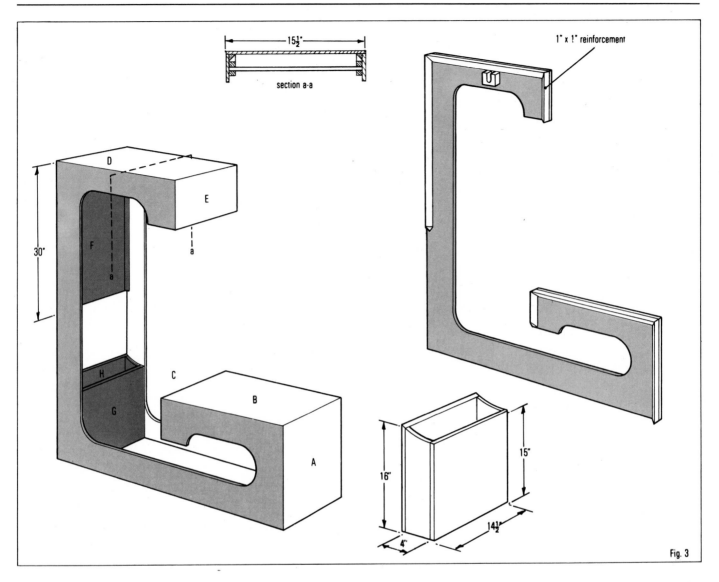

section a-a

1" x 1" reinforcement

Fig. 3

Fig 3. The construction of the hall stand is not complicated. The main construction details are shown here. The most important point to remember is that the battens at the rear of the side units must be set in $\frac{1}{2}$in (13mm) from the edge so that piece F can fit between the sides.

You can mark the curve on the top of the narrow sides of the box using the cardboard template made earlier. Cut the curve with a jigsaw with the two narrow sides held together in a vise. Smooth the curved top of the sides and then glue and nail the front and back panels G and H to the sides. At this point the whole assembly can be glued and nailed into place inside the hall stand as shown in the construction diagram in Fig. 3.

The coat rack

The coat rod for the canopy section of the hall stand is $14\frac{1}{2}$in (368mm) long. It can be made from a 1in (25mm) closet pole or 1in (25mm) tubular steel rod. This can be fixed at the ends with small blocks of plywood with U-shapes cut into them to match rod ends. These blocks can be made from cut-offs from the plywood sheet, glued and nailed in place. Alterna-

tively, you can choose one of the fittings available commercially. There are a wide variety of these and installation is not difficult. Most are just screwed in place and the rod set in place. Expandable closet rods are also available. Whatever hardware you choose, it should be secured in place about halfway along the canopy sides, and about 2in (51mm) up from the curved bottom edge of these pieces.

Finishing

Rub down the edges of the hall stand with sandpaper and trim off any irregularities with a finely set block plane. Fill any gaps at the joints with wood filler. Then paint the unit in the color of your choice. An alternate method of covering the edges is to use $\frac{1}{2}$in $\times \frac{3}{4}$in (13mm $\times$ 19mm) round molding which will have to be mitered at the corners.

Telephone table

Telephone table

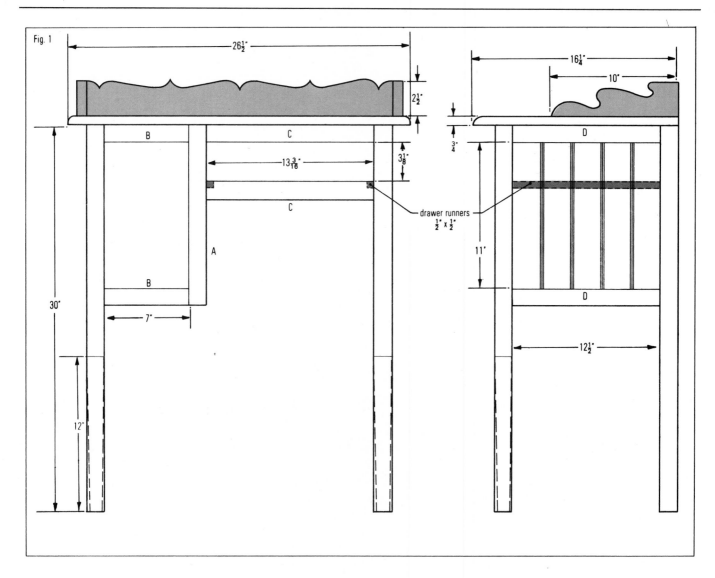

Fig. 1

A properly designed table provides a permanent, safe place for your telephone, as well as a home for those directories. In this way it will help make your home a neater, more attractive place. And making a phone call becomes much easier when the telephone can rest securely on a surface which also can hold a notepad for added convenience.

This table has been built to a traditional design, and the baroque scroll around the back and sides of the top lends a touch that would enliven a modern hallway. If you prefer a simpler style, just omit the scroll or, alternatively, alter the outline as shown to a more modern pattern.

Under the table top, one half comprises a directory storage cabinet, and the other half provides a drawer and a recess to enable a small stool to be housed – out of the way and ready for those long conversations!

Preliminaries

The joints for the legs, uprights and cross members are all stopped mortise and tenons. The mortise and tenon is a relatively simple joint but, if you prefer, you could use a doweled joint.

$\frac{1}{2}$in (13mm) plywood is used for the sides of the table. This is laid across the inside of the cross members and is glued and nailed. The top is cut and trimmed to size. Cut from a single piece of $\frac{3}{4}$in $\times$ $16\frac{1}{4}$in $\times$ $26\frac{1}{2}$in (19mm $\times$ 413mm $\times$ 673mm) pine.

A jigsaw will be required to cut the curved baroque pattern on the scroll. This is quite easy to mark if you make a cardboard template, but considerable care is needed in cutting if you don't want to end up trying to remove irregularities in the curves with a spokeshave.

The joints for the corners of the drawers are rabbeted.

Try always to purchase boards that are straight. Ones that are warped will create construction problems. Lumberyard dealers will usually allow you to select the pieces from their stock. Clean up all pieces prior to assembly as it would be difficult to obtain a good finish when the unit is completed.

Construction

Cut all four legs from $1\frac{1}{2}$in $\times$ $1\frac{1}{2}$in (38mm $\times$ 38mm) lumber to the finished length 30in (762mm). On the end-grain of one end of each piece make a mark. The marked ends indicate the tops of the legs.

Place one of the legs on a table and mark off a line 18in (459mm) from the top of the leg. Continue the line around all four sides.

On the end-grain of the bottom of the leg, mark $\frac{1}{4}$in (6mm) all around to form a square as shown in Fig. 2. Place the leg horizontally in a wood vise, with enough of it clear of the vise to enable you to plane the leg down both easily and efficiently.

With a jack plane set to fine, carefully dress down to the marked taper. Starting at the bottom of the leg remove waste and gradually work back up the leg so that the plane is cutting parallel to the marked taper. Plane the taper down to one side of the square marked on the end and repeat on the remaining three sides, and on the remaining three legs.

The side frames

First build the two side frames. The outline for these is shown in Fig. 1. Cut four cross members (D) 1in × 1½in × 14in (25mm × 38mm × 356mm) to size, then cut tenons on both ends of each side frame. The dimensions for these are shown in Fig. 3.

Lay two legs down on a flat surface, about 14in (356mm) apart, then place two cross members (D) in position in between. One member must run between the leg tops, flush with the end, and the other one below this, with 12in (305mm) in between as shown in Fig. 1. Mark off the positions of the mortises, using the tenons on the cross members as templates, then cut out the mortises in the legs. Repeat this with the remaining two legs and cross members.

When everything fits properly, glue and fit all joints, square the frames one on top of the other and hold the joints in place with two bar clamps while the glue sets.

Finish off by fitting the plywood ends between the D members of each frame. The ends of the ply are cut to fit flush with the top of the upper D member, and flush with the bottom of the lower one. When these parts are joined up, they should be glued and nailed with 1¼in (32mm) brads.

The center frame

Using the procedure described above, construct the center frame which comprises the uprights (A) 1½ × 1½ × 15in (38mm × 38mm × 381mm) the remaining two D members, and the plyboard. Note that the top mortise in piece A goes right through to receive pieces B and C while a third mortise enters at the back to receive cross piece D. The tenons may require slight trimming to make sure of an absolutely accurate fit. Therefore, it is essential that each tenon be trial-fitted before final gluing.

To ensure that the eventual fit will match, lay the right-hand frame (on the

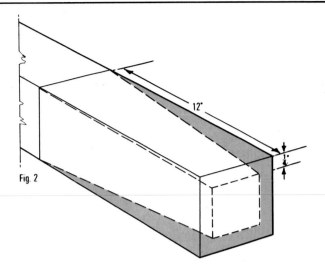

Fig. 2

Fig 1. Front and side elevations of the telephone table showing overall dimensions.

Fig 2. Detail of the legs showing the dimensions of the tapers.

Fig 3. Details of the cross member joints.

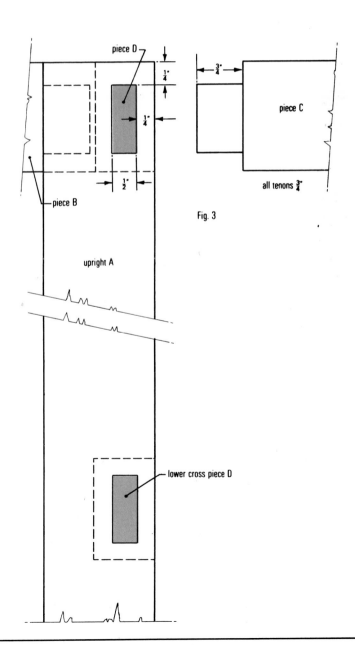

piece D

piece B

piece C

all tenons ¾"

Fig. 3

upright A

lower cross piece D

Telephone table

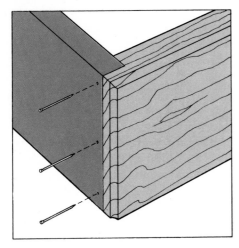

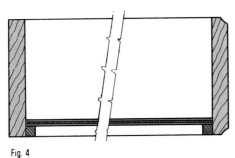

Fig. 4

left in Fig. 1) down on a flat surface, and mark and fit the center frame using the frame underneath as a guide.

The directory cabinet

The next stage is to build the directory cabinet onto the end frame.

Stand the end and center frames on their back edges on a flat surface about 7in (178mm) apart. They should be in the same positions as shown in Fig. 1, resting on what will eventually be the back of the table.

Mark and cut the four cross pieces B, 1in × 1½in × 8½in (25mm × 38mm × 216mm),then cut tenon joints in both ends of each piece as shown in Fig. 3.

Using the same procedure as that just described use the tenon joints as templates to mark out mortise joints in the legs and uprights. When this has been done, glue and clamp all joints until the glue has set. Finish off this section by cutting and fitting the plywood base of the cabinet. This must overlap the cross members and is glued and nailed in position.

Finishing the frame

When the adhesive has set properly, use the same technique to fit the four cross members(C) 1in × 1½in × 14$\frac{11}{16}$in (25mm × 38mm × 373mm) and the remaining end frame.

While the glue is setting, mark the lengths of the drawer runners, then cut them to size. The drawer runners lie between the inside ends of the lower C runners, and are simply butted, glued and nailed in place. Do this when the glue has set on the C members.

Fitting the top

Stand the frame upright, on its legs. Place the uncut table top in position, and cut and trim where necessary.

Take the top off and, with a bench plane set to fine, round off three top edges (the front and two sides) so that each edge has a similar surface.

Drill screw holes around the top frame members, place the table top back in position and drive the screws home from underneath.

Fitting the back

There will be two openings left in the rear of the frame – the cabinet and drawer spaces. Mark a line ½in (13mm) from the inner edges of openings and cut a rabbet recess ¼in (6mm) deep around each opening. This is best done with a router; clear the corners out with a sharp chisel. Note that the bottom of the directory case forms the rabbet on this side.

Cut the plywood backs to fit into these recesses, and glue and nail them in place.

The drawer

The drawer is constructed as in Fig. 4, using a router to cut the rabbets in the front and back. The front panel is chamfered ¼in (6mm) deep around the edges and 1in (25mm) onto the face of the drawer. Mark the lumber with pencil lines and using a smoothing plane set very fine, gently plane down to the chamfer lines. Glue and nail ¼in (6mm) square strips to the inside, flush with the bottom, to support the base. Fit the drawer into the table. Guides made of thin strips of wood can be glued to the runners if required.

Cutting the scrolls

These are an optional part of the structure. Mark out an outline with tracing paper, then transfer the outline to the ¾in × 2½in (19mm × 64mm) lumber.

Use a jigsaw for cutting along the outline, smoothing down any rough edges with a spokeshave or other smoothing tool.

The scroll is held in position with a good woodworking adhesive. It remains only to varnish or paint the unit, and add an attractive handle to the drawer. You will soon be wondering how on earth you ever managed without a telephone stand.

Cutting list

Solid wood

	standard	metric
4 legs	1½ × 1½ × 30	38 × 38 × 762
2 uprights (A)	1½ × 1½ × 15	38 × 38 × 381
4 cross pieces (B)	1 × 1½ × 8½	25 × 38 × 216
4 cross pieces (C)	1 × 1½ × 14$\frac{11}{16}$	25 × 38 × 373
6 cross pieces (D)	1 × 1½ × 14	25 × 38 × 356
1 table top	¾ × 16¼ × 26½	19 × 413 × 673
2 drawer runners	½ × ½ × 13	13 × 13 × 330
2 drawer base supports	¼ × ¼ × 14⅜	6 × 6 × 365
1 drawer front	¾ × 3 × 13	19 × 76 × 330
1 drawer back	⅜ × 3 × 13	10 × 76 × 330
2 drawer sides	⅜ × 3 × 15	10 × 76 × 381
1 rear scroll	¾ × 2½ × 26½	19 × 64 × 673
2 side scrolls	¾ × 2½ × 10	19 × 64 × 254

Plywood

	standard	metric
3 sides	½ × 12½ × 15	13 × 318 × 381
1 storage box base	½ × 7 × 15¼	13 × 178 × 387
1 storage box rear	¼ × 7½ × 12½	6 × 191 × 318
1 drawer base	¼ × 12¼ × 14⅜	6 × 311 × 365

You will also require:

Adhesive. Nails and screws. Paint or varnish. Turpentine or turpentine substitute.

Victorian butler's tray

Above: The Victorian butler's tray is ideal for casual meals or snacks. When not in use it can be folded up for convenient storage.

First used in the eighteenth century, these trays were used as a sideboard by the butler. Later, in the Victorian and Edwardian eras, they acted as dumbwaiters in the ritual of afternoon tea. Nowadays they are useful for holding everything from drinks to television snacks – the simple design and foldaway construction is readily adaptable to changing times.

The tray

The unit is in two parts: the tray, which has one short side cut away to allow easy access to glasses or plates; and a folding X-shaped stand on which the tray rests.

Victorian butler's tray

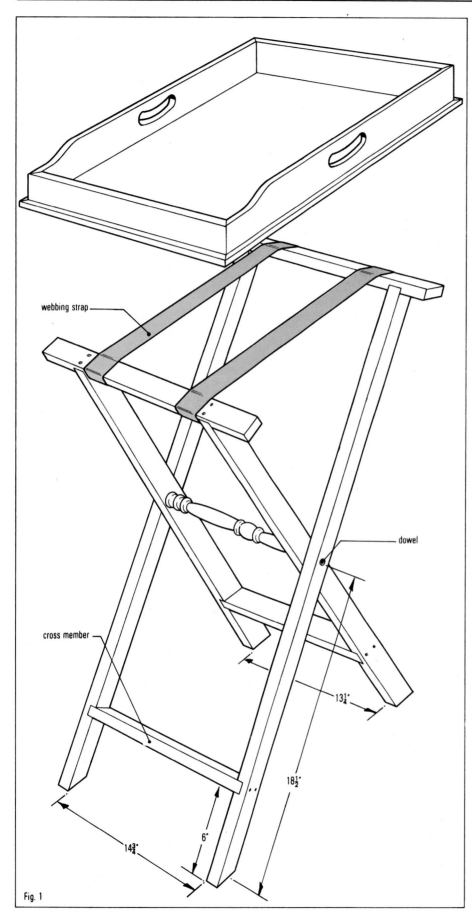

webbing strap

cross member

dowel

$13\frac{1}{4}"$

$18\frac{1}{2}"$

$6"$

$14\frac{3}{4}"$

Fig. 1

To prevent the tray slipping off the stand, the bottom is covered in felt or baize; and two canvas webbing pieces attached between the tops of the frames hold the frames open.

Choosing wood

Hardwood should be used because it does not stain as much as softwood. Traditionally, mahogany is used but there are attractive alternatives. Ideally, try to use a good quality wood possessing all the essential qualities and having an attractive finish. In this project the whole frame and the sides and tray edgings are constructed from a high quality wood, while the tray base is made from $\frac{1}{2}$in (13mm) plywood veneered with matching mahogany.

Cutting out and preliminaries

Cut out all the pieces to the sizes given in the cutting list. The two long sides are cut down at one end and handles cut in them. Figs. 2 and 3 are templates which give the exact dimensions of these features. Use a jigsaw for cutting them but do not attempt to cut exactly to the mark; you will find it easier to leave a small margin which can be sanded down later. You will need to bore a large hole in the handle opening to insert the jigsaw blade.

The top edges of the tray sides and the outer edges of the edging strips are rounded with a spokeshave.

Miter joints are used in the construction of the tray sides. Cut these joints on the ends of the side pieces and edging strips taking care they will butt together exactly.

At this stage, sand all the pieces, and if there are any irregularities on the joints they must be removed.

Assembling the tray

Begin by attaching the edging strips to the base of the tray with glue and brads. The pins should be skew nailed and punched under the surface of the wood, and the holes filled with a filler to match the wood.

The sides of the tray are attached to the base so that they overlap the edging to the base joint by $\frac{1}{8}$in (3mm) all around.

Fig 1. The butler's tray has a simple construction. The tray rests on the stand and a layer of felt backing prevents its slipping. The style of the center cross member shown may be changed to straight dowel or square lumber if you do not have access to lathe facilities.

Attach by gluing and pinning initially, and to assist in keeping the miter joints correctly aligned, clamp blocks of softwood to the internal corners of the tray.

Final attaching is done by screwing 1in (25mm) No.6 flathead wood screws through the base into the sides. Four screws on each side are sufficient; their heads must be recessed flush with the surface.

Assembling the stand

The stand is made up of two leg frames which are made separately. One leg frame is narrower than the other so that it will fit closely inside the wider frame. Make the wider frame first.

Dado joints are made at the junctions of the legs and cross members. Mark two lines $\frac{3}{4}$in (19mm) apart, 6in (153mm) from the bottom of each leg. Mark or gauge a depth line $\frac{1}{4}$in (6mm) down the edge of the leg. Cut down to this line with a crosscut saw and then remove the waste with a chisel. Ensure that the bottom of the housing recess is perfectly flat. On one cross member the housings are made $\frac{3}{4}$in (19mm) from the ends and on the other, for the smaller frame, the housings are made $1\frac{1}{2}$in (38mm) from the ends.

Assemble the frames with glue and flathead $1\frac{3}{4}$in (45mm) No.6 steel screws.

The center cross member

If you own a lathe this piece can be turned on the same pattern shown in Fig. 1. Otherwise you can fit a $1\frac{1}{2}$in (38mm) diameter piece of dowel without shaping it. Lathework is described in the Techniques section of this book.

A $\frac{3}{8}$in (10mm) diameter hole is bored $2\frac{1}{4}$in (57mm) deep into the center cross member. A corresponding $\frac{3}{8}$in (10mm) hole is bored right through the center point of the inner leg frame and $\frac{1}{2}$in (13 mm) deep into the center point of the inside of the outer frame member. Take two $3\frac{1}{2}$in (89mm) hardwood dowels and drive them through the inner leg frame into the center cross member. Slip the outer frame over the inner and locate it on the projecting dowels by 'springing' it into position. The frame will now move

Figs 2, 3. Patterns for the side and handle cutouts.

Fig 4. Section through the tray illustrating the method of construction. Care must be taken when screwing the base to the sides not to hit the nails attaching the edging strip.

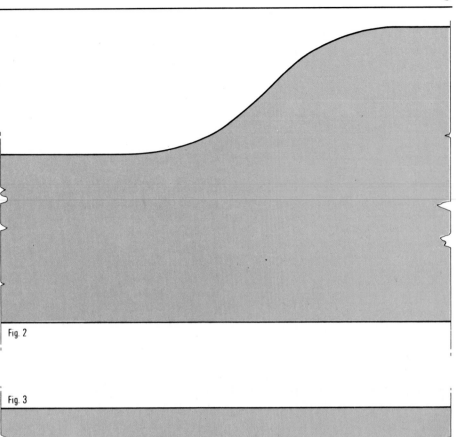

Fig. 2

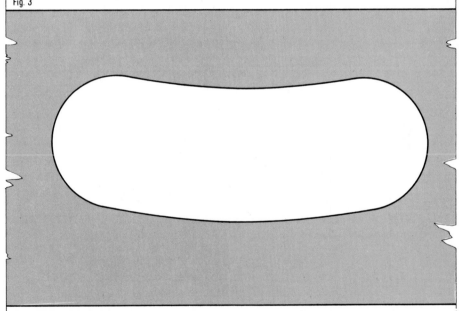

Fig. 3

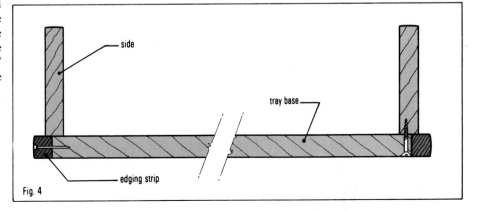

side

tray base

edging strip

Fig. 4

around the dowels which allows you to fold the stand when not in use.

Attaching the canvas straps

The stand when open should be fixed so that the angle at the center between frame members is 60°. This is achieved by fitting canvas webbing strips between the top cross members. Take a 30in (762mm) length of webbing and mark lines $4\frac{1}{2}$in (114mm) from each end. Place the webbing across the two top cross members and open the stand so that $4\frac{1}{2}$in (114mm) of webbing projects at both ends. Tack the webbing in position. The free ends of the strap are wrapped around 1in × 2in (25mm × 51mm) hardboard plates which are then screwed underneath the leg frame top cross members. Repeat the procedure with the other strap and when both are firmly secured, remove the tacks.

Levelling the legs

To give greater stability to the stand the legs must be levelled off so that their bases are flush with the floor. This involves cutting a section off each leg. The easiest method of measuring and cutting the right amount is as follows.

First, place the stand on a perfectly flat surface. If there is any wobble, caused by one leg being shorter than the others, place waste pieces of wood under the short leg until the frame stands level. Now take a waste block of wood measuring about 1in × 3in × 4in (25mm × 76mm × 102mm). Drill a hole slightly less than

the diameter of a pencil through the center of the 3in (76mm) edge. Push a pencil into this hole so that about 1in (25mm) protrudes, then lay the waste block on the flat surface so that the pencil is parallel to the floor and $1\frac{1}{2}$in (38mm) above it. Carefully draw a line around the bottom of each leg and carefully cut through the legs around these lines.

Finishing and varnishing

Sand all the surfaces for a fine finish. With a soft cloth dampened with turpentine substitute wipe all the surfaces so that the fine dust caused by sanding is picked up.

Using a good quality 2in (51mm) brush, sparingly apply a varnish of clear matt polyurethane and turpentine substitute diluted 50/50. When this coat has dried rub down with fine sandpaper, wipe the surface with the cloth again and re-coat with the same mixture. Leave this coat to dry, then finish with grade 'O' steel wool. Clean the surface with full-strength turpentine substitute, then leave overnight, and apply undiluted polyurethane as sparingly as possible.

Add the piece of baize or felt to the base of the tray. After a few days the unit can be polished with a soft, dry cloth to give it a deep rich, lasting sheen.

With only a few hours of work you will have built a butler's tray that is a perfect reproduction of a valuable antique. The simple, but strong, construction ensures that it will give years of useful service.

Cutting list
Solid wood

	standard	metric
2 tray sides	$\frac{1}{2} \times 3 \times 27$	$13 \times 76 \times 686$
1 tray end	$\frac{1}{2} \times 3 \times 20$	$13 \times 76 \times 508$
1 tray end	$\frac{1}{2} \times 1\frac{3}{4} \times 20$	$13 \times 45 \times 508$
2 tray edge strips	$\frac{1}{2} \times \frac{1}{2} \times 27\frac{3}{4}$	$13 \times 13 \times 705$
2 tray edge strips	$\frac{1}{2} \times \frac{1}{2} \times 20\frac{3}{4}$	$13 \times 13 \times 527$
1 tray base	$\frac{1}{2} \times 19\frac{3}{4} \times 26\frac{3}{4}$	$13 \times 502 \times 680$
4 stand legs	$\frac{3}{4} \times 1\frac{3}{4} \times 37$	$19 \times 45 \times 940$
2 stand top cross members	$\frac{3}{4} \times 1\frac{3}{4} \times 18$	$19 \times 45 \times 457$
1 stand lower cross member	$\frac{3}{4} \times 1\frac{3}{4} \times 14\frac{1}{4}$	$19 \times 45 \times 363$
1 stand lower cross member	$\frac{3}{4} \times 1\frac{3}{4} \times 15\frac{3}{4}$	$19 \times 45 \times 490$
1 turned center cross member	$2 \times 2 \times 13\frac{3}{8}$	$51 \times 51 \times 340$
or round center cross member	$1\frac{1}{2}$ diameter $\times 13\frac{3}{8}$	38 diameter $\times 340$
$2\frac{3}{8}$in leg pivot dowels	$3\frac{1}{2}$in length	89mm in length

You will also require:

40 brads. Wood adhesive. Finishing nails. 16 1in (25mm) No.6 flathead steel wood screws. 20 $1\frac{3}{4}$in (19mm) No.6 flathead steel screws with furniture plugs. 2 canvas webbing straps 30in × 2in (762mm × 51mm). 1 piece of baize or felt $26\frac{3}{4}$in × $19\frac{3}{4}$in (680mm × 502mm). Clear matt polyurethane. Turpentine substitute.

Coffee table

The versatile top of this coffee table fits neatly into the frame and can be lifted off and turned over.

Construction

The construction outline is shown in Fig. 1. The main point to bear in mind is that the four side rails must be the same length as the long dimensions of the table top: in this case 17¾in (455mm).

The base frame, as shown in Fig. 1, is very easy to make, and consists of four leg members joined by four rails. Two of the rails are fixed so that their top edges are below the tops of the legs at a depth equal to the thickness of the table top. The remaining rails are fixed at right angles to these, immediately underneath.

The lower rails are screwed into the leg members. The top rails are secured by screwing through the legs, into the end-grain of the rails. Screwing or nailing into end-grain does not provide a very strong joint and, for this reason, the end-grain of the top rails is drilled to take No.8 size fiber screw plugs. The plugs are glued into these holes and, when the adhesive has set, provide a strong anchoring location for the screws. A detail of this joint is shown in Fig. 1.

This top is of flakeboard covered with a laminated plastic on both faces and on the edges. But you could use lumber core plywood or even solid wood if you prefer, with any suitable covering.

Assembling the frame

First cut the four legs. The length of each leg will be the eventual height of the surface of the table. So if you require a lower, or taller, table, do this by cutting the legs to the desired length.

Next cut the rails. Use fine sandpaper to smooth down all surfaces, particularly the end-grain, because this will be

Coffee table

showing on some parts of the construction.

Now prepare the top. You may have purchased a laminated or veneered top already. If it is veneered with a particular wood on the two faces and the edges you may wish to apply different veneer to one of the surfaces. In any case the top should have its final finish before you start to screw the frame together. In this way you can make any small alterations on the rail members at this stage.

Set out the table legs by marking 1in (25mm) from the top of each leg to represent the thickness of the table top. Make another mark $1\frac{3}{4}$in (45mm) below this to indicate the top rail and another mark $1\frac{3}{4}$in (45mm) further down to indicate the lower rail. Drill two holes through the leg at the top rail marking

and use the leg as a jig to mark the corresponding holes for the screws in the rails. Glue and screw these rails to the legs.

Next, drill the lower rails for two screws at each end. Glue and screw these rails into place, keeping the ends of the rails flush with the outside edges of the legs. Sand smooth, clean with turpentine substitute and apply two coats of 50/50 clear matt polyurethane and turpentine, sanding between each. Finish with undiluted polyurethane.

Cutting list

Solid wood	standard	metric
4 legs	$\frac{7}{8} \times 1\frac{3}{4} \times 13$	$22 \times 45 \times 330$
4 rails	$\frac{7}{8} \times 1\frac{3}{4} \times 17\frac{3}{4}$	$22 \times 45 \times 455$
Flakeboard or plywood		
Table top	$1 \times 17\frac{3}{4} \times 17\frac{3}{4}$	$25 \times 455 \times 455$

You will also require:

Covering material (self-adhesive vinyl, plastic laminate or other veneers), or any other covering you may prefer. 16 $1\frac{1}{2}$in (38mm) No.8 screws with plastic screw caps. Woodworking adhesive. Varnish or paint for the legs and rails. 8 fiber plugs. Turpentine.

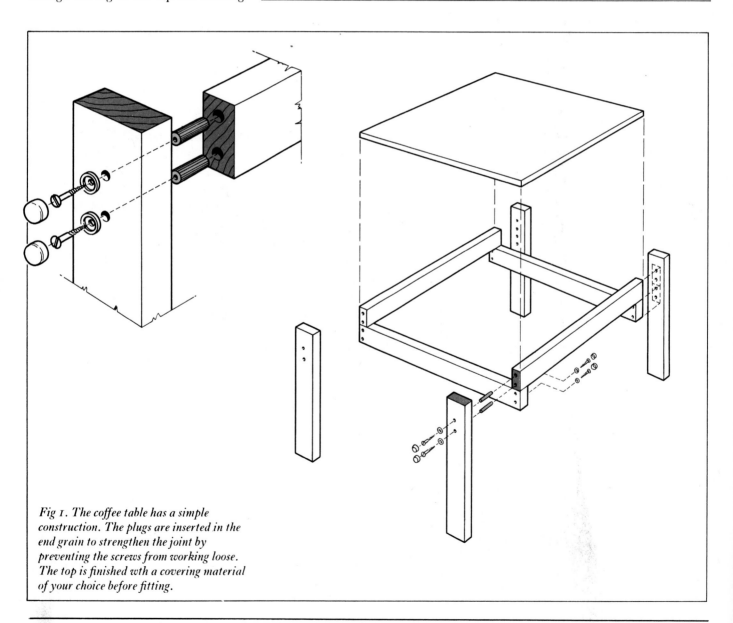

Fig 1. The coffee table has a simple construction. The plugs are inserted in the end grain to strengthen the joint by preventing the screws from working loose. The top is finished with a covering material of your choice before fitting.

Serving cart

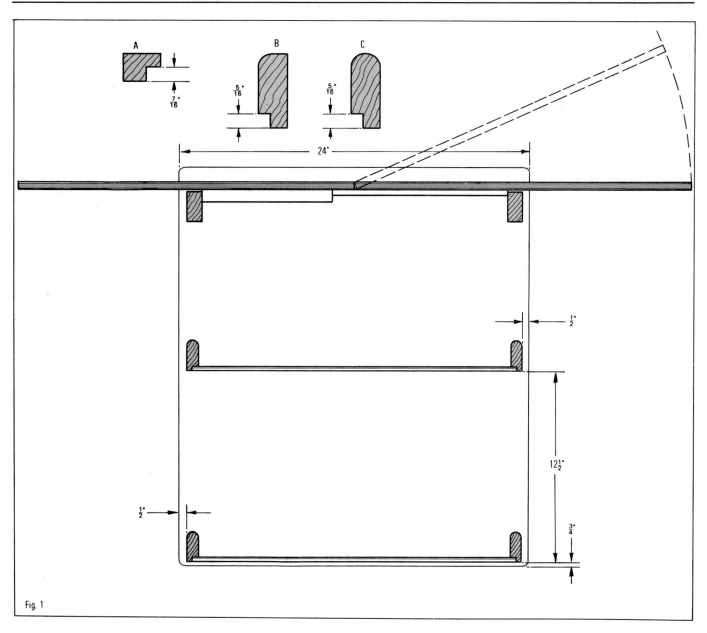

Fig. 1

Figs 1, 2. The construction details of the serving cart. The position of the center cross member is important – if it is too high the table top will be difficult to move.

As living space becomes increasingly more precious, furniture designers have concentrated on producing compact units which combine two or more functions. This serving cart certainly fits this design concept. It doubles as a trolley equipped with two easy-to-clean trays for holding food and cutlery, and a folding table top which is large enough to allow two people to eat in comfort. So, with one unique piece of furniture, you have solved two age-old problems: kitchen storage and a place for casual dining.

Construction of the serving cart reflects the straightforward and uncluttered design; there are no complicated techniques involved, and any properly equipped home handyman should be able to build the unit without encountering any problems that cannot be solved both easily and quickly.

Briefly, the trolley is made up of seven parts, consisting of the two sides, two laminated trays, the top cross members and runners, and the laminated table top. These are assembled in a strict order of working, which should be followed closely, so that the ultimate result is a truly professional one.

Preliminaries

When you have checked that you have all the necessary tools and materials, mark out all the pieces to the sizes given in the cutting list. Do not cut all the pieces out yet; greater accuracy will be achieved by working in stages, checking each piece as it is cut.

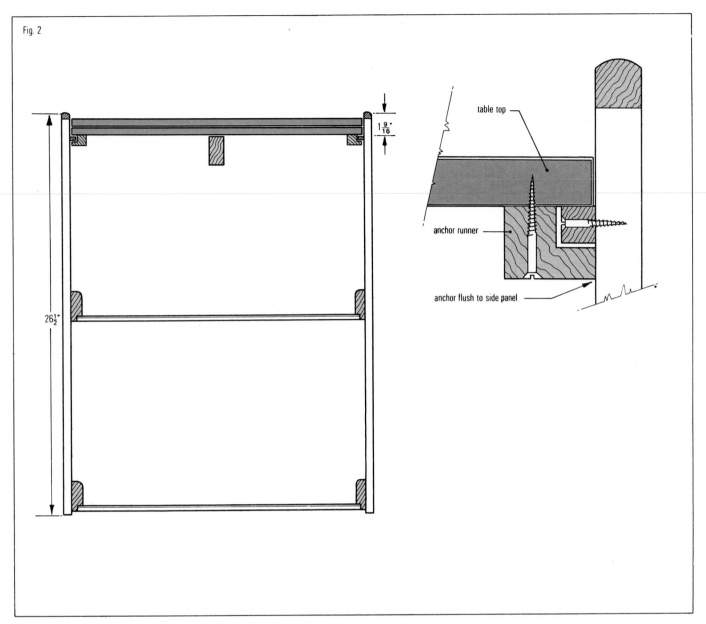

Fig. 2

table top

anchor runner

anchor flush to side panel

$26\frac{1}{2}"$

$1\frac{9}{16}"$

The cart sides
These are constructed of $\frac{1}{2}$in (13mm) plywood panels cut to size with a circular saw with a combination blade. Sand all the cut edges with medium, then fine, sandpaper and check that the edges are flat and smooth enough to take the $\frac{1}{2}$in (13mm) edging strip. Now cut the edging strip to the correct lengths. These strips are located on the top and side edges of the panels and are fixed by gluing and nailing. Using a small punch, sink the finishing nails well below the surface to allow the surfaces to be rounded at a later stage.

The tray frames
Cut the sides and ends from $\frac{3}{4}$in (19mm) softwood. All these pieces need a rabbet which houses the tray bottoms. The rabbets are $\frac{3}{8}$in (10mm) square (Figs. 1B and C), and run the whole length of each member. Rabbeting is relatively simple when done with a router, radial arm saw or table saw.

The tray side members are housed in rabbets cut in the ends of the cross members. These rabbets are $\frac{3}{4}$in (19mm) wide and $\frac{3}{8}$in (10mm) deep and can be cut out with a router. When the rabbets have been cut and tested for fit, assemble the tray frames with glue and $1\frac{1}{4}$in (32mm) screws countersunk below the surface. Now round off the entire top edges of the ends (Fig. 1C) and the internal half of the top edges of the sides (Fig. 1B).

At this stage round the edges of the cart's side panels and round off the corners. Clean up and sand all the parts and fill any holes or cracks.

Fitting the frame to the side panels
First cut out the two top main cross members and measure these against the shorter side of the tray frame to ensure that they are the same length. Then cut the top center member to size.

Refer to the section shown in Fig. 1 to mark the correct location of the trays and top cross members onto the side panels. The bottom edge of the lower tray is located $\frac{3}{4}$in (19mm) from the bottom edge of the side panel, the bottom edge of the center tray is situated $12\frac{1}{2}$in (318mm) above this point. Each top cross member is positioned $1\frac{5}{8}$in (41mm) below the top edge of the side panels.

Serving cart

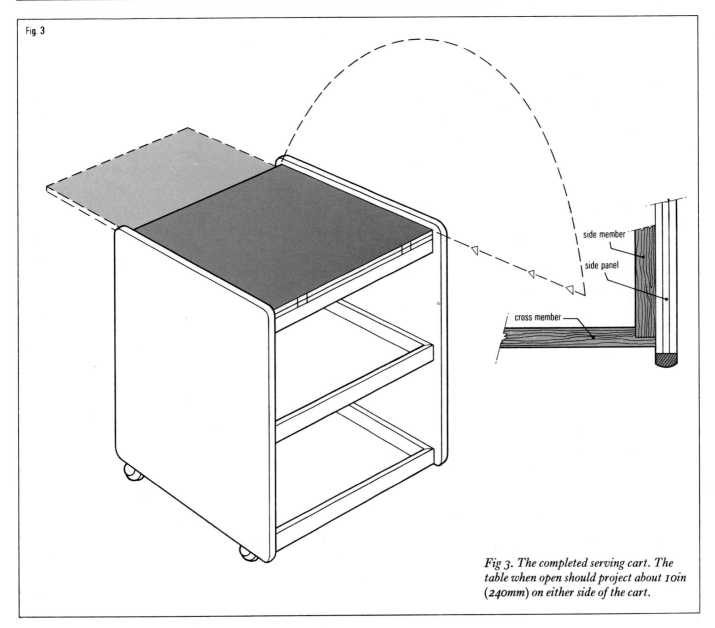

Fig. 3

side member

side panel

cross member

Fig 3. The completed serving cart. The table when open should project about 10in (240mm) on either side of the cart.

Take one side panel and secure the tray frames into position, using glue and 1in (25mm) screws inserted through the tray sides into the side panel. The top cross members are secured with glue and 1½in (38mm) nails. Then turn the structure over and secure the second panel in position in the same way.

The table top runners
The table top runs on three bearing points, the center member which has already been cut and two ⅜in (9mm) square hardwood runners which are fixed to the side panels. Cut these hardwood runners to size and fit these and the center member into position according to the plan shown in Fig. 2. All the runners must stand 1/16in (2mm) clear of the cross

members so that the sliding table top at no point comes into contact with them. The hardwood runners are fixed with glue and ¾in (19mm) screws to the side panels; the center member is glued and nailed between the cross members. At this stage sand the entire structure and fill any holes with wood filler.

The tray bottoms
The bottoms of the trays are made of hardboard covered with a rigid plastic laminate 1/16in (2mm) thick. If you intend putting hot food containers on the trays, choose a laminate that is heat resistant. Cut out the hardboard tray bottoms to size and test for fit by temporarily locating them in the rabbets cut into the bottom of the tray frames.

Now cut the laminate to the same size as the tray bottoms, using a router with a carbide-tipped bit and straightedge. You can also use a table saw with a carbide-tipped blade. The chances of chipping the material is slight with these tools.

Ensure that the rough sides of the hardboard tray bottoms are dry and grease-free, then apply a thin but even layer of contact adhesive onto them.

In the same way, apply the adhesive to the back of the laminate and wait until the adhesive on both surfaces is nearly dry before pressing the two surfaces together. Clamp the two materials together to give a strong bond, pounding them together with your hand.

When the adhesive has set properly, drill No.4 countersunk screw holes at

5in (127mm) intervals around the tray bottom. The center of each screw hole is situated $\frac{3}{16}$in (6mm) from each edge to coincide with the center of the rabbet. The tray bottoms will be secured at a later stage in construction – do not do this now.

The table top

This is constructed from two $\frac{1}{2}$in (13mm) panels entirely covered by laminate and hinged together. The size of these panels is given in the cutting list but, due to errors in cutting the other pieces, the measurements may have to be modified very slightly. In order to construct a well-fitting table top, first measure the width between the tops of the side panels, then subtract $\frac{1}{4}$in (6mm) from this measurement to give the exact width of the table top. The subtraction represents $\frac{1}{16}$in (2mm) clearance on each side of the table between table and side panel, and the thickness of the two laminate edging strips.

Mark out the correct dimensions of the table tops and cut these pieces out. Now mark out and cut the laminate sheets and edging strips, remembering to allow $\frac{1}{8}$in (3mm) excess all around which can be trimmed down later. One other way to cut the laminate is to lay a straightedge in place along the line to be cut and, using a utility knife or straight laminate cutter, start scoring gently from one end of the sheet. Hold the cutting blade against the straightedge to keep it straight while it travels the entire length of the pencilled guideline. Do this at least three or four times, increasing the pressure of cut each time, until the dark under-surface of the laminate appears as a clean, unbroken line.

Work steadily and do not rush, whatever method you use. Laminate, while extremely hard and durable, is also relatively brittle and it can be cracked or chipped fairly easily. Since it is expensive material it will definitely pay to go slowly. If the straightedge is solid (a length of particleboard with a laminate edging is most suitable for this purpose), you should encounter no difficulty in the cutting operation.

When all the laminate has been cut to size, glue the shorter edging strips to the shorter edges of the table and use a belt sander to sand these down flush with the table top and sides before adding the long edging strips. Cut these to size, then cover the laminated surfaces using the

technique detailed previously for the trays. Hinge the two flaps together with brass hinges, after first having cut out the recesses in the table tops to the required depth, thereby allowing the hinges to be properly flushed.

Attaching the table to the frame

Lay the folded table top on a flat, clean surface and pack it up $\frac{1}{2}$in (13mm) clear of the surface. Now turn the frame upside down and place it on the table top, and carefully align it all around to give the correct location. One flap of the table top is secured to the frame by anchors which slide along the two outer hardwood runners. Cut these anchors to the size given in the cutting list and make a $\frac{7}{16}$in (11mm) square rabbet in them as shown in Fig. 1A. This rabbet loosely houses the hardwood runner and so, when the anchor runner is fixed to one table flap, the whole table top slides easily along the runners.

To attach the anchors when the table top and frame are correctly aligned, lay them on the table flap with the rabbet enclosing the runners. Having checked that there is sufficient clearance between the rabbets and runners, locate the long edges of the anchors almost flush against the inside faces of the side panels, with their ends butting against one of the top main cross members. Once in position they are screwed to the table with 1in (25mm) screws whose heads should be

recessed slightly just below the surface.

Turn the whole structure the correct way up and test the operation of the table top. The closed flaps should slide to project approximately 10in (254mm) beyond the main structure and the hinged table top should then turn over to project approximately 10in (254mm) on the other side. If there are large discrepancies in these measurements, or if the table top does not slide easily, adjust the runner anchors.

Painting the cart

It is easier to paint the cart before fixing the tray bottoms in place and with the table top removed. It is a simple operation to unscrew the anchor runners and the job of painting will be made much simpler. Make sure to apply a coat of primer to the surface before one, if not two, undercoats, allowing each coat to dry thoroughly before applying the next. Finally, finish with a coat of a dirt and heat-resistant paint, giving you an attractive and extremely durable addition to your home.

When the paint has thoroughly dried, re-attach the table top and add the tray bottoms by screwing them into the rabbet housings in the tray frames. Attach ball-type castors to the base of the cart by screwing them onto small hardboard plates glued to the corners of the lower tray base. Finally, lubricate all the runners with candlewax to ensure that the table top slides easily.

Cutting list

Solid wood	standard	metric
4 softwood tray ends	$\frac{3}{4} \times 2 \times 22$	$19 \times 51 \times 559$
4 softwood tray sides	$\frac{3}{4} \times 2 \times 22\frac{1}{2}$	$19 \times 51 \times 572$
2 softwood top main cross members	$1 \times 2 \times 20$	$25 \times 51 \times 508$
1 softwood top center member	$1 \times 2 \times 21\frac{1}{4}$	$25 \times 51 \times 540$
2 hardwood table top runners	$\frac{3}{8} \times \frac{3}{8} \times 21\frac{1}{4}$	$10 \times 10 \times 540$
2 hardwood table top runner anchors	$\frac{3}{4} \times 1 \times 10$	$19 \times 25 \times 254$
4 softwood edging strips	$\frac{1}{2} \times \frac{1}{2} \times 26$	$13 \times 13 \times 660$
2 softwood edging strips	$\frac{1}{2} \times \frac{1}{2} \times 24$	$13 \times 13 \times 610$
Plywood		
2 plywood panels	$\frac{1}{2} \times 24 \times 26$	$13 \times 610 \times 660$
2 plywood table flaps	$\frac{1}{2} \times 19\frac{7}{8} \times 23$	$13 \times 505 \times 584$
Hardboard		
2 hardwood tray bottoms	$\frac{1}{4} \times 19\frac{1}{2} \times 22\frac{1}{2}$	$6 \times 495 \times 572$
Laminate		
4 sheets plastic laminate	$\frac{1}{16} \times 20\frac{1}{8} \times 23\frac{1}{4}$	$1.5 \times 510 \times 590$
2 sheets plastic laminate	$\frac{1}{16} \times 19\frac{1}{2} \times 22\frac{1}{2}$	$1.5 \times 495 \times 572$
1 strip plastic laminate	$\frac{1}{16} \times \frac{3}{4} \times 192$	$1.5 \times 19 \times 4877$

You will also require:

2 1in (25mm) hinges, 4 mini ball-type castors. Wood adhesive, contact adhesive. 60 1$\frac{1}{4}$in (32mm) finishing nails, 48 1$\frac{1}{4}$in (32 mm) wood screws.

Sofa bed

The design of the couch has been kept as simple as possible and this is reflected in the wide choice of materials available for the construction. Depending on how much you are prepared to spend and the particular finish you desire, you can use flakeboard, plywood, softwood or hardwood for the basic frame – or any combination of these materials. For example, you could use laminated plywood for the sides, back, base and fascia, and softwood for the base supporting frame. In this way, you reproduce the design economically, and the finished couch lends itself well to a fine paint finish. Or, you can use solid wood throughout to produce a more solid and sophisticated result. This latter method lends itself to further modification. The sides can be built of single wood panels or can be made up of separate strips of wood joined together, as shown in Fig. 3. It is this construction which is described here, but the basic construction method is the same whatever materials you use.

There are several different ways of making the cushion-supporting base. Perhaps the simplest is to fit a single plywood panel across the frame. Provided you supply adequate cushioning, the result will be perfectly comfortable. Another, but more expensive, method is to fit rubber webbing across the frame.

Store-bought cushions are expensive and it is unlikely that you will be able to obtain them in a size to fit the couch. By following the instructions given in this project you can easily make up the necessary cushioning yourself.

Making the sides

Having decided on what materials to use, begin by cutting out the side panels; or, if you intend making up the side panels from strips of lumber, cut each piece to the size given in the cutting list. Each side consists of four parallel members with a cross-section of $1\frac{3}{4}$in × $5\frac{1}{4}$in (45mm × 133mm), which are flanked by two vertical members of the same section. Fig. 5 shows how the side and top members are mitered together for greater strength, and also shows the tongue-and-groove joints by which the separate pieces are fixed to the sides.

To make these joints, first take each of the vertical side members and, having marked in a line from one corner, at an angle of 45° to the long edge, cut the mitered end. In the same way, miter the ends of the top horizontal member, but remember that this piece has a tongue cut on it and, therefore, has to be cut a minimum of 1in (25mm) overlength at each end.

Now take the vertical members and make a groove 1in (25mm) deep and $\frac{1}{2}$in (13mm) wide down each miter and the inside long edge. Clean up these grooves with a chisel and sandpaper. Then take the top horizontal member and, with a rabbet plane, cut a tongue on each mitered end 1in (25mm) long and $\frac{1}{2}$in (13mm) wide. Similarly, make tongues on each short edge of the other horizontal members as shown in Fig. 5.

If you have no rabbet plane, you can cut the tongues in the ends of the lumber like cutting tenons. First saw down the grain to make a central $\frac{1}{2}$in (13mm) tenon, then cut across the grain to form the shoulders of the joint.

With the pieces of both side panels cut to shape, trial assemble and, when you are satisfied that each joint fits perfectly, begin construction by gluing the lower horizontal members into the grooves on the vertical members. Complete the construction by adding the top in the same way. Allow the glue to set, then trim off that part of the miter-tongue which protrudes beyond the outer edge of the vertical members.

Sides and back assembly

Assembling the side panels to the back is straightforward. However, you must ensure that the butt joints between side and back panels are completely secure, and this cannot be achieved by simply screwing through the side panels into the end-grain of the back panel. One way to attach them is with dowels. Or, you can bore the short edges of the back panel at 4in (102mm) centers to receive fiber plugs. Fit the sides into position and screw through them, at previously marked locations, using 3in (76mm) wood screws countersunk below the surface. The surface should be filled with a suitable filler. A special countersink bit can be fitted to your drill.

Sofa bed

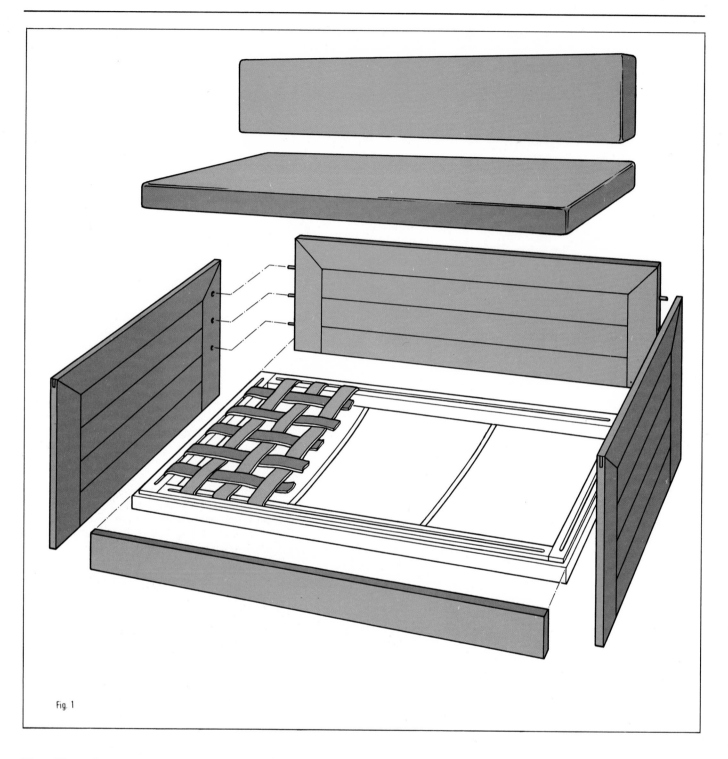

Fig. 1

Fig 1. The exploded diagram of the sofa bed shows the main components.

Figs 2-6. Details of the construction. The conduit is very important as it strengthens the frame. It is these two components which take the stress when the unit is in use.

Making the base frame

The frame which supports the seat is constructed separately, then fitted, pre-assembled, to the back and sides. To make it, begin by cutting the rails to the sizes given in the cutting list, then miter all the ends to an angle of 45°. Ensure that each mitered end joint is smooth, then glue and nail the rails together, checking that the structure is square by measuring the diagonals. A fascia panel is

fitted across the front rail, but this piece is not added until the base of the seat has been fixed to the frame. It is then glued and screwed to the front rail of the seat and the side frames.

Next, fit the frame inside the side and back panels, with its lower edges located 5¼in (133mm) above the floor. Fasten the frame with 3in (76mm) wood screws driven through the frame members at 6in (152mm) intervals. The frame must

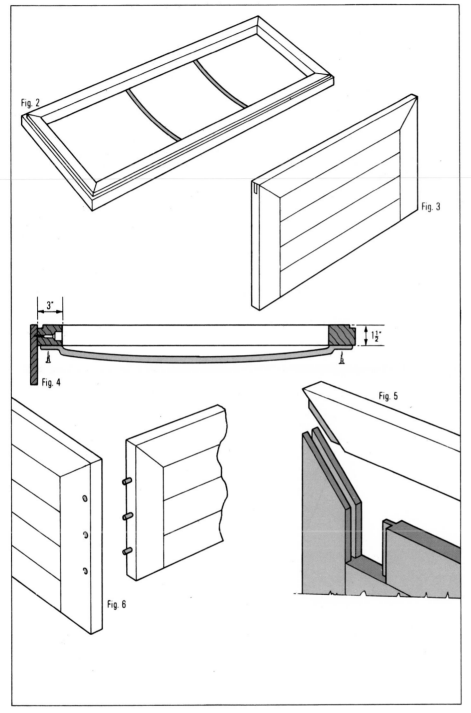

Fig. 2

Fig. 3

Fig. 4

Fig. 5

Fig. 6

flatten the ends with a hammer, drill the ends to receive screws, then screw the tubing across the underside of the frame at the locations shown in Fig. 2.

Making the base

As mentioned above, you can make the base from a single sheet of laminated plywood ½in (13mm) thick. This is a good strong material. Simply cut the base panel to the correct size and screw onto the top edges of the frame.

The only disadvantage of plywood as a material for the base of the couch is its rigidity. It will not yield under pressure and, consequently, you must make cushions extra thick. An alternative, more luxurious base is provided by plastic webbing stretched between the frame sides. Webbing can be obtained in a continuous roll and there are several methods of fixing it in place. The easiest way is simply to tack or staple it to the top edges of the frame, having first removed any sharp edges with which the plastic may come in contact. Another method is to attach clips to the ends of the webbing, then anchor the clips in a groove cut in the top of the frame. Whatever method you choose to fix the webbing, make sure that it is installed under the correct tension and will give complete support.

Making the cushions

Both the seat and the back cushions use foam rubber as a foundation. This is both inexpensive and easy to cut and the only problem occurs in choosing a foam filling of the correct density. It is this density which determines the resilience of the finished cushion. As the unit will be used for sleeping on, you must take extra care in choosing and making up the foundation.

This will depend, to some degree, on how firm you want the foundation to be. It is best to seek the advice of an upholstery supplies dealer.

You have a choice between natural and synthetic foam. The former is, of course, more expensive but will last longer. Different foams also have different densities.

Cut the foam to size, using a fine-toothed hacksaw or an electric carving knife. All the foundations should be cut slightly oversize; the reason for this being that when they are covered by fabric of the correct size, they will be slightly compressed, preventing wrinkles forming in the fabric cover.

be pre-drilled to a depth of 1½in (38mm) in order to make starting holes for the screws in the side panels.

Because the couch is long and must bear the weight of more than one person, reinforcement members must be fixed across the frame to maintain its strength and rigidity. Conduit tubing cut and bent to the shape shown in Fig. 4 is ideal. To make these pieces, cut ¾in (19mm) tubing to the lengths given in the cutting list,

Sofa bed

Right: An alternative method of construction. The dowels for the fascia panel are allowed to show as a decorative feature.

Cutting list

Solid wood	standard	metric
2 top side members	$1\frac{3}{4}\times5\frac{1}{4}\times34\frac{1}{2}$	$45\times133\times869$
8 horizontal side members	$1\frac{3}{4}\times5\frac{1}{4}\times24$	$45\times133\times610$
4 vertical members	$1\frac{3}{4}\times5\frac{1}{4}\times26\frac{1}{4}$	$45\times133\times666$
3 horizontal back members	$1\frac{3}{4}\times5\frac{1}{4}\times66\frac{1}{2}$	$45\times133\times1682$
1 top back member	$1\frac{3}{4}\times5\frac{1}{4}\times75$	$45\times133\times1905$
2 vertical back members	$1\frac{3}{4}\times5\frac{1}{4}\times21$	$45\times133\times533$
2 frame cross rails	$1\frac{1}{2}\times3\times29\frac{1}{2}$	$38\times76\times756$
2 frame long rails	$1\frac{1}{2}\times3\times75$	$38\times76\times1905$
1 fascia panel	$1\times6\times75$	$25\times152\times1905$

The size of the couch can be modified to suit individual requirements. For economy, man-made wood, such as plywood, can be substituted for solid wood.

You will also require:

Plastic webbing or a plywood base. 3 lengths $\frac{3}{4}$in $\times29\frac{1}{2}$in (19mm $\times$ 749mm) tubing. Foam cushion foundations and fabric covers. Wood glue. 12 fiber plugs. 3in (76mm) wood screws. Wood filler. Sandpaper. Wood stain and varnish, or paint.

Studio couch & chair

Studio couch & chair

These two units comprise a studio couch and chair, built to the same design. The couch 6ft 9in (2.1m) long and the chair 2ft 6½in (763mm) can be built as individual units or as a complete living room suite. The couch is perfectly adequate for sleeping adults, while the chair can convert to a small bed suitable for a child.

The unit consists of a basic frame, which is the support structure, and two additional frames for the seat and backrest. When you wish to convert the unit for sleeping, two bolts, which secure the backrest to the rear legs, are withdrawn. This allows the hinged backrest to fall into a horizontal position, supported at the far end by two retractable legs as in Fig. 2.

Although the backrest is fully upholstered, the retaining bolts and legs, situated in the interior of the backrest, are easily reached through an opening.

The type of lumber
The lumber used in this couch is African mahogany, a hardwood suitable for furniture, flooring, veneers, joinery and construction work. Other hardwoods could also be used. You may prefer something a little lighter, or darker, or with a richer grain. Or, other woods may be dictated by your budget.

Tools and materials
In addition to commonly used tools you will require a bit for your power drill, capable of drilling a hole through a carriage bolt and nut.

For the studio chair, you will need two 6ft (2m) bar or pipe clamps and, if a couch is being made, two extension bars will be needed for the clamps. It is not usually necessary to go to the expense of buying bar clamps, as they can often be rented. Ask at your local hardware or do-it-yourself store.

There are several items of hardware you will need: four ⅜in (9.5mm) carriage bolts, 4in (100mm) in length and washers shown in Fig. 8; used to cover the carriage bolt recesses; four hinges for the couch, two for the chair, each 2½in (64mm); two 2in (50mm) casement screws (Fig. 9) to lock the backrest.

The method of upholstering will be detailed later. But it is a mistake to do more than provisionally decide on a color scheme and material at this stage.

Buying the lumber
Hardwood for furniture should be purchased from a lumber dealer who specializes in hardwoods. You will require 'prepared' lumber – which is supplied machined and planed in exact widths, but only approximate lenghts. So, when you give your order, give the exact measurements for the widths, but add an inch (25mm) to each length illustrated in Figs. 3 to 5 just to be certain.

Before you order the lumber inquire whether your dealer will exchange any pieces that have unsightly blemishes. Most reputable lumber dealers will agree to this.

Marking out
Before attempting to mark out the lumber read through the instructions with reference to the drawings in order to familiarize yourself with the design and function of each part.

Using the try square and marking knife, mark off the exact length of each piece of wood. As each piece is marked, carefully examine each surface to ensure that the best face on each piece will be placed where it is visible. It is pointless, for instance, having the most attractive surface of an armrest on the underside, where it will be out of sight.

Mark out the joints. The rails are joined to the legs by stopped mortise and tenon joints. To avoid weakening the legs, the mortise and tenon of each side rail are 'haunched'. In this way they will not meet the front and rear tail tenons at right angles. The arms are joined at the rear legs by twin mortise and tenon, and at the front legs by open twin mortise and tenon. The joint at each corner of the seat and back carcasses is formed by a box or comb joint. Details for the making of these joints can be found in the Techniques section.

For the plywood panels, which support the cushions, a grid of 3in (76mm) squares is drawn on the surface of each panel. This will provide a network of 'crosses' created by lines meeting at right angles. Each cross is drilled with a brace and ¾in (19mm) bit to form the rows of holes as illustrated.

Cutting the joints
It is best to cut out all the joints before anything is fitted. Note that some of the box joints are mitered at a 45° angle. These are at 1, the bottom of the front edges of the backrest frame panels and 2, the top of the back edges of the seat frame panels. These miters will allow the backrest to be brought up to a seating position (Fig. 5).

Mark the tenons on the rails and cut them out. Then mark the legs using the tenons as a guide. Cut the mortises using the measurements given in Figs. 4 and 5. Most of the waste can be drilled out with a suitable sized bit and the mortise hole can then be squared up with a chisel. Do not forget that the haunch only goes into the leg ½in (13mm).

The open mortises for the chair arms and the combed joints for the seat and back carcasses are cut with a tenon saw and then the waste is chopped out with a chisel working from each side of the wood to give a neat finish and avoid splitting the lumber faces. When marking out the combed joint shade the waste pieces with a pencil, as it is very easy to start cutting on the wrong line.

Cut out the two blocks which are used to house the center brace of the seat frame. Do not at this stage drill any holes for the bolts or screws.

Pre-assembly finishing
The final finish of the woodwork is a matter of personal preference. But whatever method you choose, there will be surfaces like the underside of the arms, for example, that are almost impossible to treat once the couch has been built. All such surfaces should be finished at this stage.

Assembling the frame
The first parts to assemble are the sides of the main frame. Each comprises a front and rear leg, an armrest, and a side rail. Each side should be glued and clamped, one on top of the other to ensure that they are exactly the same, and left for the glue to set.

While the sides are setting, the seat and backrest frames could also be glued, providing you have additional bar clamps. If not, you might be able to improvise by wedging, as shown in Fig. 6. The unit to be clamped is placed between two battens screwed to the bench, and wood wedges are driven between one batten and the unit to force the clamping action. If you have a really solid bench you could get away with placing one end of the frame against a wall and using only one batten or block.

When the glue has set on the sides of the main frame, the front and rear rails (the 'long' rails if you are making a couch) are glued and fitted to the sides, and clamped. The bar clamp will be adequate for a chair, but you will need extension bars for the clamp if you are making a couch.

When the glue has set on the back and seat frames, fit all interior battens. Glue these and screw them at 2in (51mm) intervals.

Glue and screw the blocks for the center brace (or blocks for both braces for a couch) and then glue the center brace into the block slots.

At this point you will have to cut recesses at the front of the front panel of the seat frame just behind the box joint (Fig. 1). This is necessary in order to accommodate a double thickness of upholstery. Using the crosscut saw, cut to a depth of about $\frac{1}{4}$in (6mm) just inside each joint. Mark lines across the width of the panel $1\frac{1}{2}$in (35mm) in from the ends of the panel. Chisel down from these lines to the bottom of the $\frac{1}{4}$in (6mm) cut.

Chisel recesses to house the hinges. Fit and screw the hinges for the junction of the back and seat frames. Place the seat frame in position. This will enable you to mark and drill the carriage bolt holes in the correct positions. Insert the carriage bolts, but don't screw them in firmly.

Raise the backrest up to seating position. Drill the holes for the wing screws. Fit and secure the screws.

The folding legs are now fitted to the inside side panels of the backrest.

Drill the holes in the plywood panels, plane the panels to fit, and glue and screw them to the interior battens of the seat and backrest frames.

The studio couch or chair is now fully assembled ready for the upholstery.

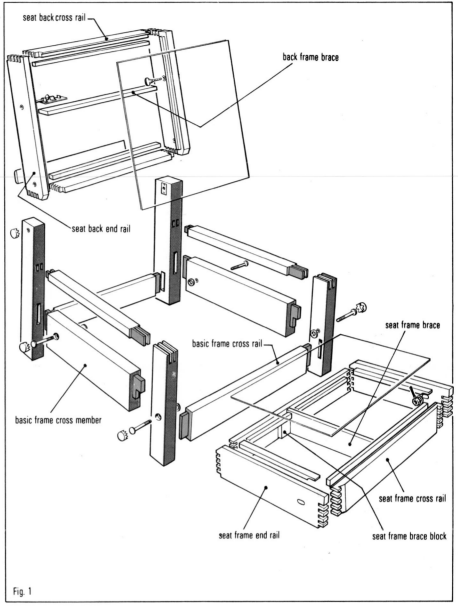

Fig. 1

Upholstering

For upholstering the couch you will need a 6in × 30in × 75in (152mm × 762mm × 1905mm) foam for the seat and a 6in × 24in × 75in (152mm × 610mm × 1905mm) one for the back; to pad the seat and the back, two strips each of 1in × 2$\frac{1}{2}$in × 75in (25mm × 64mm × 1905mm) and $\frac{1}{2}$in × 10in × 75in (13mm × 254mm × 1905mm).

For the chair you will need a 6in × 24in × 24in (152mm × 610mm × 610mm) foam for the back, and a 6in × 24in × 30in (152mm × 610mm × 762mm) one for the seat; for padding, two strips each of 1in × 2$\frac{1}{2}$in × 24in (25mm × 64mm × 610mm) and $\frac{1}{2}$in × 10in × 24in (13mm × 254mm × 610mm).

The foam slabs should be of medium hard density and the padding of hard foam. These are available at upholstery shops, or see your Yellow Pages.

Other materials you will need are: No.3 or 4 piping cord, contact adhesive (make certain that it is a kind that will not dissolve foam), decorative upholstery tacks and gimp pins.

Cushion covers

If you are using plain fabric allow a piece of furnishing fabric (minimum 48in [1220mm] wide) equal to twice the length of each foam biscuit, plus 2in (51mm) seam allowance (the box strips can be cut from the leftover fabric of the main pieces).

If using a fabric with a one-way design, you will have to make up the length for the cushion cover by joining widths of fabric so that the pattern will run the

Fig 1. The component parts for the couch and chair. The seat center brace and blocks are omitted from the construction of the chair.

right way on both the backrest and the seat cushion, i.e. from top to bottom for the back, and back to front for the seat. To determine the number of widths of fabric you need double the length of each foam slab to give the amount of fabric needed for the top and bottom sections of the cover.

Follow the same method to calculate the amount needed for the box strips on each side of the foam, multiplying the total of fabric widths by the cushion's depth plus 1in seam allowance.

To cover the fascia and back, allow 2yd (1.8m) of 48 in (1220mm) wide fabric for the couch and 1yd (91.4cm) for the chair. For the amount of bias-cut casing fabric and piping, measure the perimeter of each biscuit and double it, allowing an extra 6in (157mm) for joining. For the fascia piping, you will need an additional 18in (457mm). (As a guide 1yd of 48in [91.4cm of 1220mm] fabric will make about 28yd [25.6m] bias strip, $1\frac{1}{2}$in [38mm] wide.)

Cutting out and making up
Cut out the fabric for the top and bottom sections of the cover, leaving $\frac{1}{2}$in (13mm) seam allowance, and join the widths for each side if using one-way patterned fabric. Cut the box strips for each side separately, again leaving $\frac{1}{2}$in (13mm) seam allowance on all sides. Join all the box strips together along their short edges with a $\frac{1}{2}$in (13mm) plain seam, tapering the stitching into the corners $\frac{1}{2}$in (13mm) from both ends. Make the casing for the piping and attach it to both sides of the box strip so that the line of stitching is $\frac{1}{2}$in (13mm) from the edges.

With the 'right' sides together, pin one edge of the box strip to the outer edge of the cover top. Clip the casing at the corners – the tapered seam of the strip will give enough 'ease' to go around the corners smoothly, so there will be no need to clip this. Baste and machine stitch and pink the raw edges if the fabric is likely to fray. Press.

Attach the bottom cover to the other edge of the box strip in a similar way, but leave one of the long sides open. Turn the cover right-side out and press it. Insert the foam and slip stitch the opening together, using a curved needle (these stitches can be unpicked easily for removing the cover for cleaning).

Covering the fascia and back
For the padding, use contact adhesive to

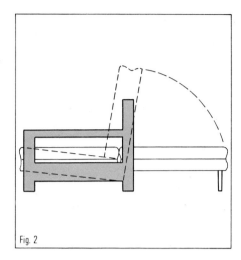
Fig. 2

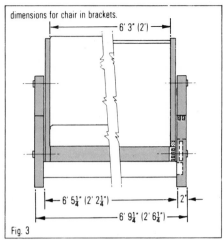
dimensions for chair in brackets.
Fig. 3

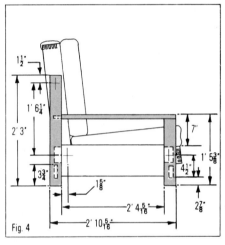
Fig. 4

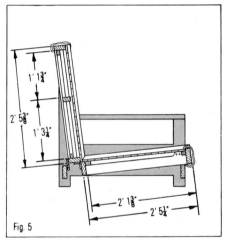
Fig. 5

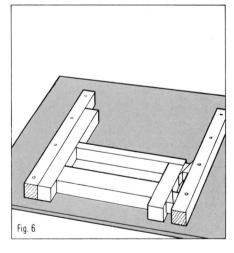
Fig. 6

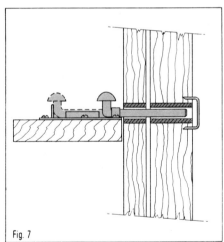
Fig. 7

Figs 2–5. Overall dimensions and main construction details for the studio couch.

Fig 6. Method of clamping the basic frames while gluing.

Fig 7. The seat back locking mechanisms may be cut from ordinary door bolts.

For the back, cut a piece of fabric the same size as the $\frac{1}{2}$in (13mm) foam plus 1in (25mm) all around. Then cut another piece to fit the inset section of the back, plus $\frac{1}{2}$in (13mm) seam, leaving the ends open 8in-10in (203mm-254mm) to provide either for the insertion of zippers, or hooks and eyes. Turn under the outside edges $\frac{1}{2}$in (13mm) and press down. Baste the fabric over the foam.

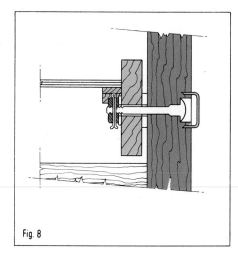

Fig. 8

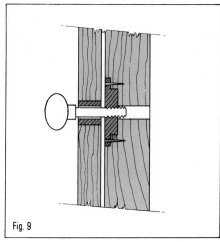

Fig. 9

Fig 8. The pivot for the seat. Note the cotter pin through the nut and bolt to prevent the assembly coming undone.

Fig 9. As an alternative to the door catch, a casement screw can be used. This is a better arrangement, which will give a more professional finish.

stick on the strips of 1in (25mm) thick foam to the front edge of the seat and the top of the back. Wrap the $\frac{1}{2}$in (25mm) thick foam around these and the seat or behind the back, and stick them in position.

For the fascia, cut out a strip of fabric the same size as the $\frac{1}{2}$in (25mm) foam, plus 1in on all sides. Stitch piping to the short sides and then turn under the $\frac{1}{4}$in (6mm) seam allowance and piping casing, and press. Fix the fabric over the foam, using plain or decorative-headed upholstery tacks. To secure the sides, lift up the piping and fasten with gimp pins.

Cutting list: Couch

Solid wood	standard	metric
Basic frame		
2 legs	$2 \times 3 \times 17\frac{3}{8}$	$51 \times 76 \times 442$
2 legs	$2 \times 3 \times 27$	$51 \times 76 \times 686$
2 arms	$1\frac{3}{4} \times 3 \times 33\frac{5}{16}$	$45 \times 76 \times 840$
2 cross members	$2 \times 6\frac{1}{2} \times 32\frac{13}{16}$	$51 \times 165 \times 834$
2 cross rails	$1\frac{1}{4} \times 4 \times 80\frac{1}{4}$	$31 \times 102 \times 2038$
Seat frame		
2 cross rails	$1 \times 4 \times 77$	$25 \times 102 \times 1956$
2 end rails	$1 \times 5 \times 29\frac{1}{4}$	$25 \times 127 \times 743$
1 front interior batten	$1 \times 2 \times 75$	$25 \times 52 \times 1905$
1 back interior batten	$1 \times 1 \times 75$	$25 \times 25 \times 1905$
1 end interior batten	$1 \times 1 \times 25\frac{1}{4}$	$25 \times 25 \times 641$
1 center brace	$1 \times 3\frac{5}{8} \times 27\frac{3}{4}$	$25 \times 92 \times 708$
4 center brace blocks	$1 \times 3\frac{5}{8} \times 3\frac{5}{8}$	$25 \times 92 \times 92$
Back frame		
2 cross rails	$1 \times 4 \times 77$	$25 \times 102 \times 1956$
2 end rails	$1 \times 5 \times 29\frac{3}{8}$	$25 \times 127 \times 745$
1 center brace	$1 \times 3 \times 75$	$25 \times 76 \times 1905$
2 interior battens	$1 \times 1 \times 73$	$25 \times 25 \times 1854$
1 interior end batten	$1 \times 1 \times 27\frac{3}{8}$	$25 \times 25 \times 699$
Plywood		
Seat frame	$\frac{3}{8} \times 27\frac{1}{2} \times 75$	$10 \times 699 \times 1905$
Back frame	$\frac{3}{8} \times 27\frac{3}{8} \times 75$	$10 \times 698 \times 1905$

Cutting list: Chair

Solid wood	standard	metric
Basic frame		
2 legs	$2 \times 3 \times 17\frac{3}{8}$	$51 \times 76 \times 442$
2 legs	$2 \times 3 \times 27$	$51 \times 76 \times 686$
2 arms	$1\frac{3}{4} \times 3 \times 33\frac{5}{16}$	$45 \times 76 \times 840$
2 cross members	$2 \times 6\frac{1}{2} \times 32\frac{13}{16}$	$51 \times 165 \times 834$
2 cross rails	$1\frac{1}{4} \times 4 \times 29\frac{1}{4}$	$32 \times 102 \times 743$
Seat frame		
2 cross rails	$1 \times 4 \times 26$	$25 \times 102 \times 660$
2 end rails	$1 \times 5 \times 29\frac{1}{4}$	$25 \times 127 \times 743$
1 front interior batten	$1 \times 2 \times 24$	$25 \times 51 \times 610$
1 back interior batten	$1 \times 1 \times 24$	$25 \times 25 \times 610$
2 end interior battens	$1 \times 1 \times 25\frac{1}{4}$	$25 \times 25 \times 641$
Back frame		
2 cross rails	$1 \times 4 \times 26$	$25 \times 102 \times 660$
2 end rails	$1 \times 5 \times 29\frac{3}{8}$	$25 \times 127 \times 745$
1 center brace	$1 \times 3 \times 24$	$25 \times 76 \times 610$
2 interior battens	$1 \times 1 \times 22\frac{3}{8}$	$25 \times 25 \times 569$
2 end interior battens	$1 \times 1 \times 27\frac{3}{8}$	$25 \times 25 \times 698$
Plywood		
Seat frame	$\frac{3}{8} \times 24 \times 27\frac{1}{2}$	$10 \times 610 \times 699$
Back frame	$\frac{3}{8} \times 24 \times 27\frac{3}{8}$	$10 \times 610 \times 698$

You will also require:

For couch and chair—2 2in (51mm) casement screws or door bolts. 4 carriage bolts, nuts and washers $\frac{3}{8}$in $\times$ 4in (10mm $\times$ 102mm). 4 2$\frac{1}{2}$in (64mm) back flaps. Screws. Nails.

Mix~and~match living room suite

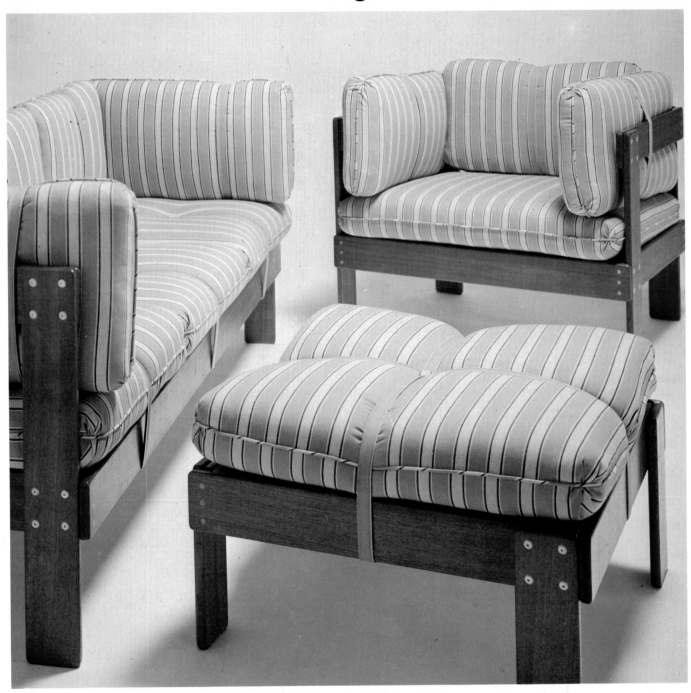

Above: Three of the suite units. The staggered legs enable the stools (but not the chairs) to be placed together along a wall.

These units are constructed mainly of utile. This is a medium-priced African wood, similar to mahogany and is suitable for most joinery projects. You may, however, prefer a lighter or darker wood, or one with a richer grain, in which case most hardwoods will do. If you have no experience in the selection of various woods, talk the matter over with your lumber dealer because the correct wood for a specific job depends not only on appearance but also price, type, and how it will eventually be finished.

Basically the main frame consists of 1in × 4in (25mm × 102mm) hardwood that has been rounded off. This should be carefully done; if it isn't it can easily result in uneven edges, and for this reason it is advisable to make a radius template, as in Fig. 1, to obtain the correct curvature. Use a router with a carbide bit to cut.

Where frame members cross by lapping, they are joined with four bolts as in Fig. 5. The positioning of the holes is extremely critical and it is essential that a hole template be made as in Fig. 2. This will ensure that drill holes match perfectly.

Where main frame members butt against one another, they are joined with three dowel rods drilled and fitted in dovetail fashion as in Fig. 3. Here again, the positioning of the holes – in this case to take the dowel rods – is extremely critical and it is essential to use a dowel jig (Fig. 3) if the members are to be joined level and at right angles. If you do not wish to make your own jig you can buy a steel jig specially made to be used with a power drill.

The method of upholstering the suite is detailed at the end of the project.

Hardware

These units are secured or joined with special fixings. The bolts are hexagonal countersunk and the nuts pronged 'T' types. The screws are Pozidriv No.8 Twinfast.

Some of this hardware may not be readily available, in all localities, but there are substitutes. You could also use chrome-plated oval head No.10 machine screws with 'T' nuts or, use carriage bolts and screws. (See parts list for sizes.) And where bolts or screws are not attractive their heads can be hidden with small domes, usually nickelled or chromed, with prongs which allow you to fasten them to the wood. There are other alternatives, but the hardware should show.

Making a radius template

This aid is used to make sure that the rounded edges of the members are uniformly curved. When held over the rounded edge, any irregularities can be seen. Note the template in Fig. 1 is drilled out with a 1¼in (32mm) bit, making the diameter slightly wider than the wood being worked. This is because only the corners are being rounded, and this is achieved with a template curve slightly larger than the thickness of the wood. If you wanted to make a template for a completely rounded edge, the drilled hole would have to be the same diameter as the wood's thickness.

It is important that the wood you use be really square and that any jagged edges inside the hole be sanded smooth.

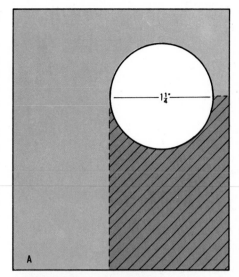

Fig. 1

Figs 1A, 1B. The pattern for the radius template.

Left: Method of using the dowelling jig.

Below left: Detail of leg and rail assembly showing both bolts and dowels.

Mix~and~match living room suite

A hole template

The holes to take securing bolts must be drilled with precision, and for this reason a template must be made to ensure that the positions of the bolt holes for the legs and rails will match.

To make the template you will require two pieces of hardboard or thin plywood, at least 2in (51mm) longer and wider than a 'long' leg member; sufficient 1in × 1in (25mm × 25mm) battening to run around three edges as in Fig. 2; finishing nails and adhesive.

Lay one piece of hardboard on a flat surface and place one of the long legs on it. Cut and fit three 1in × 1in (25mm × 25 mm) lengths of batten around the two short edges and one long side as in Fig. 2. Glue and nail the battening in position. Place the other piece of hardboard over the battening and glue and nail to complete the template body.

You now have an 'envelope' into which long or short leg members, or the end of a rail, will fit. Trim the hardboard flush with the outside edges of the battening.

The next step is to mark out two sets of holes. First ensure that the 'open' side of the template is absolutely flush with the edge of a long leg member placed inside it, so that accurate measurements can be taken.

Mark out the position of the holes as illustrated in Fig. 2, then carefully drill the holes. If you do not wish to repeat the marking process you can lay a sheet of paper over the first set of marks, outline the hole positions, and transfer the paper along, as you work, to position the next set of holes.

It is absolutely essential that the holes are drilled at right angles to the surface of the hardboard. If possible, a drill stand should be used for the job. If this is not possible align the drill bit with a try square.

Making a dowelling jig

A dowelling jig must be made to ensure that the dowel rods are accurately and consistently placed. You can see the complete dowel jig assembly shown in the illustration in Fig. 3.

The base is constructed from plywood about ¾in (19mm) thick and 30in (762mm) square, and 1in × 2in (25mm × 51mm) battening. It is essential that the support marked 'A' in Fig. 3 is of the same thickness as the leg or rail members of the furniture.

Cut the battens to length and mark out

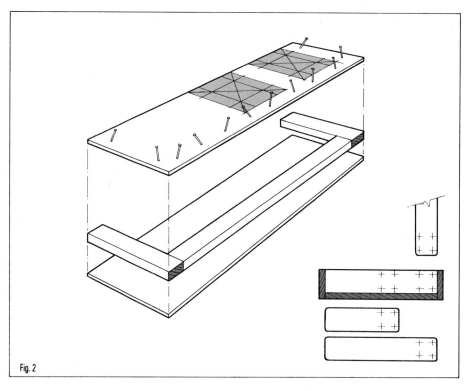

Fig 2. The drill hole template and its method of use.

their positions on the base board as detailed in Fig. 3. Glue and nail the battens in position.

The guide block 'B' correctly positions the overlap of the rails 'E', and also guides the drill bit for the dowel holes. This block should be made from hardwood – the harder the better – and is hinged to block 'D' with a 2in (51mm) back-flap hinge. This allows accumulated sawdust to be removed from the jig between drillings. When the jig is being used it should be screwed or clamped to the working surface.

Having completed the bolt hole template and the dowelling jig, you can start on the construction of the furniture. First decide whether you want to build the 'formal' suite or the 'dovetail' suite the construction outline of which is in Fig. 8.

Place the lengths of stock for assembling one unit of furniture to one side, and examine these for 'matching'. This is done to ensure that all lumber surfaces that can easily be seen – such as the outside faces of legs or rails – harmonize with one another. For instance, the leg members might come with a prominent grain on one side, and little or no grain on the other, which could present a jumbled visual appearance if fitted at random.

Mark the 'inside' of such members lightly with a pencil to identify them.

Cutting

Cut all lengths and right angles accurately to their final dimensions. This is easily achieved by nailing three lengths of batten around the edges of a leg so that it forms a three-sided stop. Next construct a 'bridge' that will exactly span the leg from side to side; this will require two short pieces of batten, and one piece sufficiently long to span the width of the leg and be nailed securely onto the two shorter battens. The whole assembly can be made in a few minutes. The bridge is held in place by a clamp at the correct distance along the leg, rail or arm, and will enable you to cut several lengths of lumber to the same length while guiding the tenon saw blade at right angles.

Using a try square and marking knife, mark one leg, rail (and arm, if required), to the correct length. Do this carefully to ensure that you do not make a mistake – you can measure as many times as you like, but you can cut only once!

The cutting should be done with a fine-toothed tenon saw with at least 14 teeth to the inch.

Rounding off (hand tool method)

Rounding the corners of the members eliminates sharp edges. It is done here to soften the lines of the furniture which, by

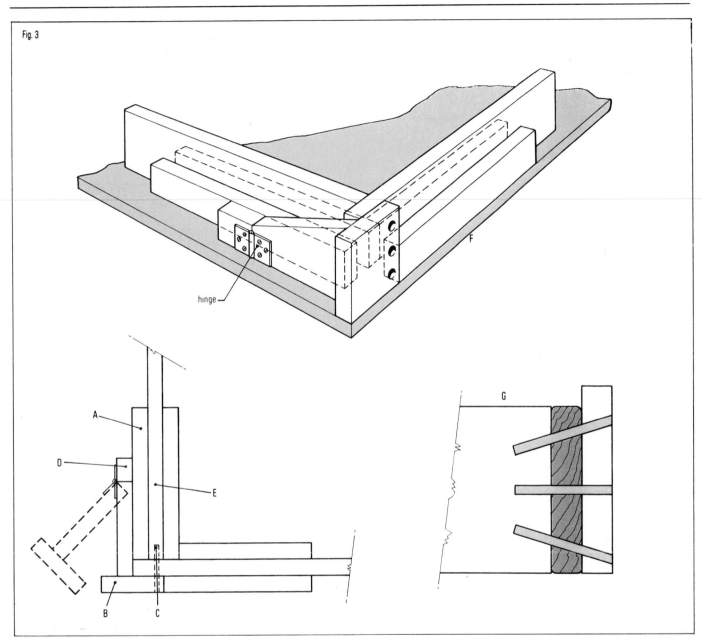

Fig. 3

A
D
E
B
C
hinge
F
G

Below: The construction of the jig, which is used to make sure the legs are square.

Fig 3. The cross rail dowelling jig. Again, care is required to ensure a good fit.

design, tends to be angular in appearance. The radius, in this case, is carried out to the extent that only the corners are rounded off, not the complete edge.

While it is a simple technique, rounding off requires a certain amount of skill and care if you are to avoid bumpy, undulating edges. If you have not done this before you should either get your lumber dealer to supply the lumber ready-cut – this will cost more and not all dealers will do it – or practice on some spare pieces of wood until you get the hang of it.

To round by hand you will need a marking gauge, $\frac{1}{2}$in (13mm) bevel-edged chisel, hammer, smoothing plane, block plane, fine-toothed flat file, and grade 'O' sandpaper. (Of course, you can also use a router.)

Set the marking gauge at $\frac{1}{4}$in (6mm) and run a *light* line at this depth along the edge of each piece of lumber, including the ends. Repeat this by drawing another line, in between the first line and the edge, setting the marking gauge at $\frac{5}{32}$in (2mm). You will now have two parallel lines marked all around the edges of the lumber.

Place one piece of lumber in a wood vise so that one long edge is facing up. If you do not have a vise, the piece can be clamped to a table top (Fig. 4); but in this case the edge will be placed horizontally,

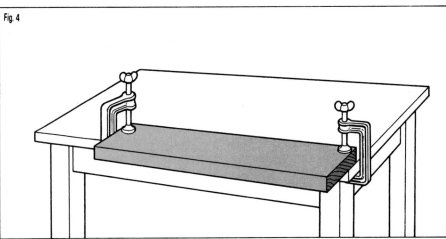

Fig. 4

Above: Leg and rail corners must be rounded using a block plane or Surform.

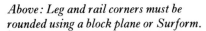

Fig 4. Each component must be clamped to a bench for rounding.

which is slightly more inconvenient as you will have to turn the lumber over to radius the other edge.

With the smoothing plane, chamfer the corners of one long edge down to the $\frac{5}{32}$in (4mm) line. Do not forget to plane with the grain, not against it, and make sure that your strokes are long and even to prevent the radius from being irregularly shaped and bumpy. Repeat this along the corner of the opposite edge. You should now have a piece with flat bevels on both edges.

Now, still using the smoothing plane, round the same corners down to the $\frac{1}{4}$in (6mm) line. Use a very light, smooth stroke for this because if you plane deeper than the line you will have to end up rounding the edges completely.

When you have rounded the long edges of all the lumber finish off the work with the fine sandpaper.

Next, you will have to round the ends, which is slightly more difficult because you are working with end grain, and this is liable to splinter or split much more easily.

Lay the length of lumber down on the bench or table top and, with the hammer and chisel, take a triangle of wood off the tip of each end corner down to the first marked line.

With the block plane, chamfer all around the end, down to the first line. As you are working on end grain, do not plane along the line as you did with the side edges because the wood will only splinter. Plane starting from each end

toward the middle. Check with the template that the curvature is uniform. If it is, continue with the block plane, chamfering down to the second line all around. After a final check with the template, sand down to a fine finish.

Rounding edges (power tool way)

To do this it's best to use a router equipped with a carbide bit and a ball bearing guide. The ball bearing guide serves two purposes. First, it acts as a guide: it ensures that the cutter slices away material at the proper depth. Second, it ensures that no burning of the wood will take place, as can happen when a cutter alone is used. These burn marks then have to be removed – a time consuming process. It's best, when working with the router, to start at the middle of the board and work completely around the board in one direction and end where you started.

Applying a protective coat

At this stage it is best to apply a protective coating so that any stains or marks that occur as the result of working can easily be wiped off. For economy, dilute the coating. A varnish consisting of clear polyurethane and turpentine or turpentine substitute in the proportions of 1:1 is perfectly adequate. Rub off all pencil and finger marks first, and wait for the coating to dry thoroughly before starting work again.

Drilling bolt holes

Mark out the positions of all the bolt holes with something that is easily erased, such as chalk. Now run through a 'trial assembly', to ensure that the bolt holes are in the correct positions. When you are sure that all the pieces have been

marked in the correct places, put each piece in turn in the drill template and mark each drill hole position positively, with an awl or a drill bit, through the template holes.

Drill the holes for the bolts through each marked position. It is essential that the holes are drilled through at exactly right angles and for this you will need a drill press. It is possible to drill the holes by hand.

Next, bore the 'T' nut and screw cup recesses. The depths to which these recesses are taken will depend on the type of bolt and nut used. And, of course, the countersink bit, in each case, should be of a diameter equal to either the nut or cup, whichever is being used.

Dowelling preparation

The dowel holes are drilled and the dowels inserted in dovetail fashion, as in Fig. 3G. Once again, mark the intended dowel hole positions with chalk and 'trial assemble' the unit. It will be even easier this time because you can insert the bolts – tightening them gently by hand – to hold part of the unit together.

Insert each piece of lumber in turn in the dowelling jig and drill holes through the outside members only. Note that the internal holes – which in all cases are drilled in the edge or end grain – are drilled when the unit is being assembled. If you drill both external and internal dowel holes together, it is almost certain that they will not match the bolt holes exactly when assembling, which will necessitate some alterations.

The dowelling rods must now be slightly chamfered. This consists of planing a 'flat' along the length of the rods. The chamfer allows air and excess glue to be forced out of the dowel holes when the dowel is being inserted. If this is not done, it will be impossible to insert the dowel to the full depth because of the excess glue.

Cut the dowelling into lengths of 2¾in (70mm). Then, to facilitate the insertion of each dowel, radius or 'round' one end. The method is the same as sharpening a pencil and can be carried out with a trimming knife or block plane. These dowels – called glue dowels – are also available ready-made.

Assembly

Place a dab of adhesive in all the screw cup holes and insert the screw cups. Insert the four bolts through the two members for one of the joints and tighten down with your fingers the pronged 'T' nuts on the ends of the bolts. Then screw them in fully. Repeat this for all the lapped joints. You will now have several parts of the unit ready, and these have to be butt-jointed with the dowelling.

Position two of these parts so that they are ready for drilling the dowel holes. This is quite simple because they will be at right angles to one another. Now clamp the corner together as shown in Fig. 6; the near end of the clamp, pressing on the surface containing the dowel holes, will be fixed directly over the middle dowel hole. Remember to use waste blocks under the clamp, otherwise unsightly dents will be left in the wood. When the assembly has been clamped firmly, place the drill bit in one of the exposed dowel holes and drill through into the second member. Repeat this on the other hole. You have now drilled the holes for the dowels that actually 'dovetail' toward the middle dowel. Smear some adhesive around the dowels, and put some of it into the dowel holes. Tap the two dowels in position. There will be a small portion of dowel protruding but ignore it at this stage – it will be planed down later.

Having completed the dowelling for the outside holes, you can remove the clamp. Now drill and fit the dowel for the center hole. This procedure is repeated for all the dowelled joints.

When the glue has set, fix the support battens around the inside of the frame-

Fig 5A. Exploded view of the chair unit.

Fig 5B. Exploded view of the corner unit of the suite. The seats of these units must be mounted level in contrast to the elevated front edge of the main seat units.

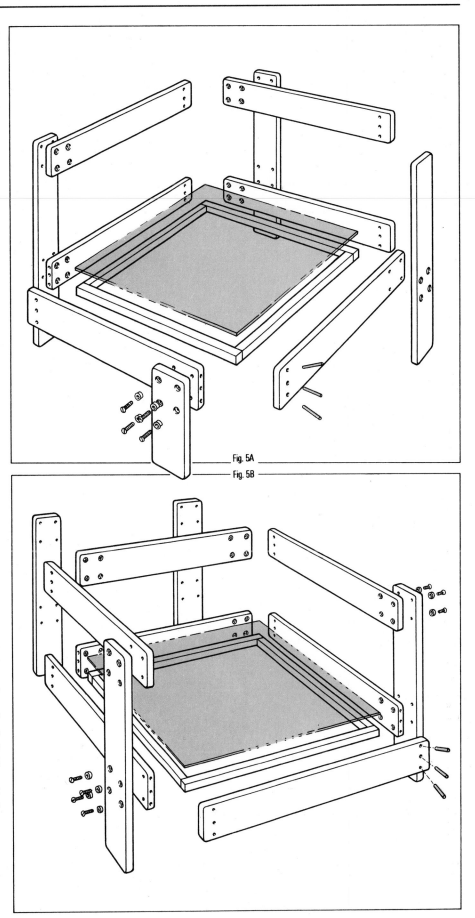

Fig. 5A

Fig. 5B

Mix-and-match living room suite

work to take the seat base plywood. For the conventional chairs these battens are positioned at an angle on the side rails from $\frac{3}{8}$in (9mm) below the front edge to $\frac{1}{2}$in (13mm) above the bottom edge of the back rail as shown in Fig. 9. For the corner chairs the battens are parallel with the rails. Drill and countersink the battens, then glue and screw them into place.

The battens for the table top must be fixed exactly to the thickness of the table top below the top of the rails. You can either measure the position, or you can cut the top to size, placing it face down on a flat surface. Turn the unit frame upside down placing this over the table top; the whole unit will now be upside down, and the top of the table will be level with the top rim of the table frame. You can now measure and fit the battening from underneath.

Finishing

Plane the protruding dowelling down carefully to prevent marking the wood and finish off with sandpaper.

Smooth all the remaining surfaces with grade 'O' sandpaper and wipe all over with a cloth dampened with turpentine to remove any dust. Then go over the wood with fine steel wool and wipe again with the cloth.

For the final finish use wax and polish, a polyurethane varnish, or paint. Allow varnish or paint to dry thoroughly.

Upholstery

The final touch is added to your suite with the construction of seat, back and side cushions for upholstery. The style of the suite is such that it blends with most modern fabric patterns.

The cushions for the suite can be made in two ways. If Dacron filling is wrapped around the foam shape before covering, you will end up with deep, pliant cushions. This is the type shown in the photographs. But if the Dacron is omitted and the foam used on its own, the cushions will be firmer, with more pronounced edges – and cheaper to make.

Materials

It is important that the foam shape for the cushions is of the correct type. A foam cushion for seating should be firmer, or denser, at the inner core with a softer exterior. The foam used here consists of a layer of hard foam sandwiched between two layers of a slightly softer foam.

Fig 6. The corners of the units must be clamped and the outer dowels inserted first. When these are dry the clamps are removed and the remaining dowel installed.

Fig 7. Detail of the construction of the corners and a plan view of the suite unit illustrating the staggered legs.

Fig 8. Plan view of the completed suite.

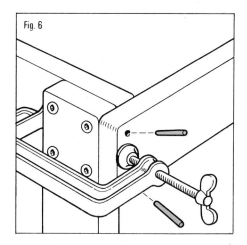

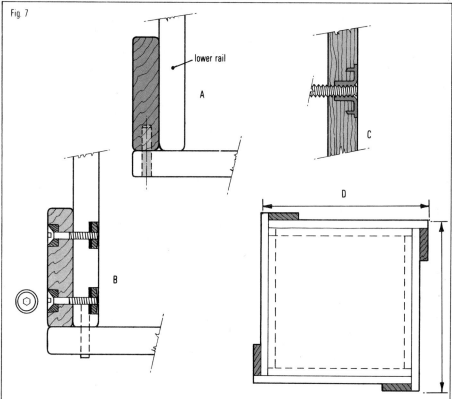

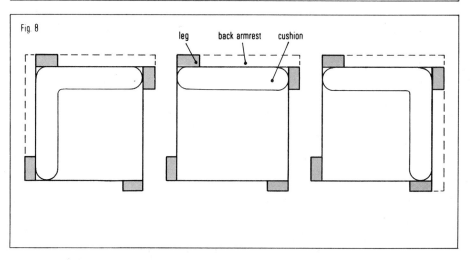

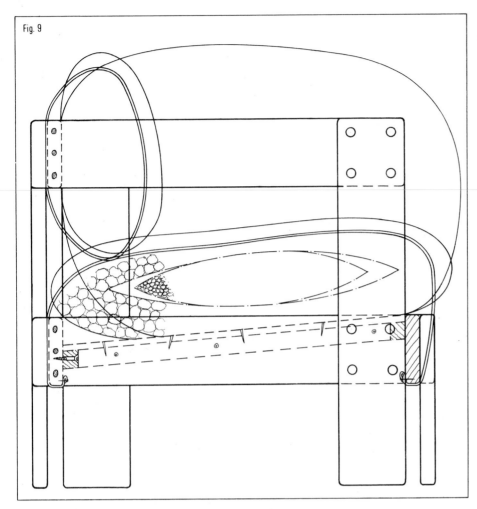

Fig. 9

Fig 9. Side elevation of the chair. Note that the battens supporting the seat must be fixed 1in (25mm) above the bottom of the rails at the back. The method of attaching the seat panel is also shown – the canted nails provide a better grip on the surrounding batten.

pieces of hardboard overlapping the edge of a table or workbench as shown in Fig. 10. The hardboard must be wide enough to spread any pressure evenly over the foam. Short strips of hardboard will not allow you to hold the foam securely over the whole area, and will tend to make you exert too much pressure at the edge, which will result in concave edges after being cut. This effect is shown in Fig. 11.

If a curved or irregular shape has to be cut, the pieces of hardboard should be tied with the foam in the center as in Fig. 12. This will enable you to handle the complete 'sandwich' as a unit instead of three separate pieces which keep drifting out of place.

A felt pen is ideal for marking the surfaces of the foam, and the best tool for cutting it is an electric carving knife. Failing this a hacksaw can be used – with or without the saw frame. If you use a hacksaw frame, ensure that the blade is set at 90° to the frame.

Shaping the edge is easily done by spreading contact ('impact') adhesive along the edges, then folding the ends toward the middle as shown in Fig. 14.

Before you do this, check with your supplier to ensure that the adhesive you use will have no adverse effect on the foam. Some adhesives will dissolve certain foams.

Padding the foam

If the foam is to be covered with filling, to estimate the quantity required, measure the length and width of a cushion and add the thickness of the foam to each measurement. In this case, your 24in × 24in × 3in (610mm × 610mm × 76mm) foam will require filling 27in × 27in (686mm × 686mm) and you will need two such pieces for each cushion. Lay the pieces along each side of the foam with the edges slightly overlapping the edges of the foam and then sew it along the join.

Making the cushion cover

For the length, measure the cushion length along the inside of the frame. To

The thickness of the foam depends on which type of cushion you are making. If you are adding a Dacron filling, then foam 3in (76mm) thick will be quite adequate; but if you are not using any additional filling, the foam must be around 4¼in (106mm) thick.

The Dacron filling is made specifically for upholstery and is sold in sheets, something like a giant roll of sterile cotton. It can be obtained from many large stores.

Piping cord will be required for running a piping seam around the joins. Buy size 1½ or 2; you will need 8ft (2.4m) per cushion, plus 10% margin.

The cushions are mainly held in place by straps, which can be made from the same material as the main covering or from any strapping or webbing that you think will match or contrast well.

Buttons are used on some of the upholstery shown. These can be ornamental or plain buttons chosen for a specific effect; or they can be covered with the main covering material. If you wish the buttons to be covered, buy the bases and cover them yourself, or ask in the store where you purchase the buttons if they can arrange to have this done.

The backrest and side cushions should be fastened to the frame – especially if you are going to omit the straps – to stop them from slipping out of place. There are several methods of doing this, such as heavy-duty snap fasteners, or Velcro tape.

A final covering can be made from any type of fabric. But it is best to use a soft fabric such as velvet or linen, if you are covering the foam with Dacron; if not adding Dacron, you should use a firmer material such as denim.

Preparing the foam

If the foam shapes are to be covered with Dacron filling, then they will need only to be cut into squares or rectangles of a suitable size. But if the Dacron is not being used, the edges of the foam should be rounded in order to prevent the appearance from being chunky and angular.

Cutting should be done between two

this measurement add 4in (102mm) to allow for the swell of the cushion and an extra 1in (25mm) for the seams. For the depth, measure the depth of the frame and add 4in (102mm) plus 1in (25mm) as before. This will give you one panel. You will obviously need two panels like this for each cushion.

Lay the fabric out on a table. Sometimes the selvage will cause the edges of the material to wrinkle or pucker, in which case the selvage should be carefully cut away. This will enable the material to lie flat.

Make sure that the bottom edge of the material is square at the corners. This is ideally done with a tailor's square, but any similar instrument such as a try square could be used.

Mark out – preferably with a yardstick – one of the panels, using tailor's chalk. Make sure, if a patterned fabric is being used, that the pattern is the right way up and in the center of the pieces to be cut.

Figs 10-14. The method of cutting the foam for the cushions. The cutting should be done between two pieces of hardboard overlapping the edge of a table or workbench.

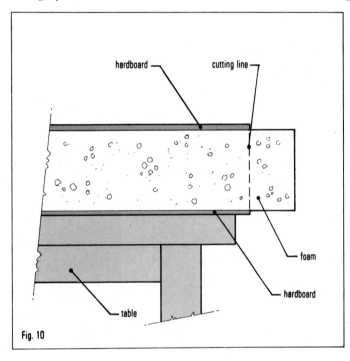

Fig. 10

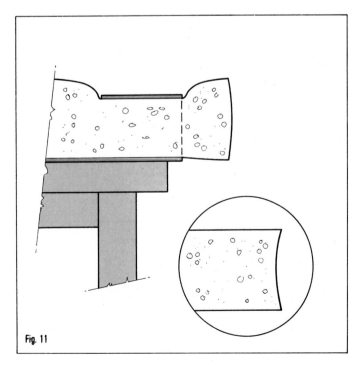

Fig. 11

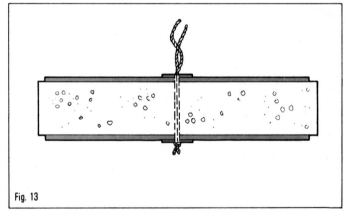

Fig. 12

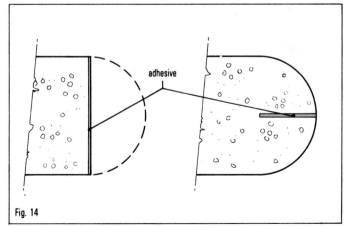

Fig. 13

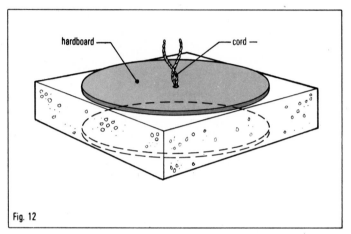

Fig. 14

Lay one panel on the table. Starting from the top left-hand corner, measure 2in (51mm) inward along the horizontal edge and mark with chalk. Repeat this down the vertical edge. Draw a chalk line between these marks and cut along the line with scissors. This will snip off the corner of the panel. Each corner of both panels must be cut off in the same way.

These corners are now stitched with a running stitch (or a long stitch if using a sewing machine) to form the gathered corners of the suite cushion covers.

Piping will have to be done on a sewing machine. Cut $\frac{1}{4}$in (6mm) strips of material on the bias (diagonally) across the material. Join these pieces end to end using $\frac{1}{2}$in (13mm) seams. Press the seams with an iron.

Fold the piping strips evenly, placing piping cord inside the fold. Sew the cord into the piping strip (using a piping foot or one-sided foot on the machine) until you have enough piping to go around all

Below: The suite may be finished in either clear or colored polyurethane. The cushions are attached underneath with snap fasteners.

Mix-and-match living room suite

four sides of one of the panels. Sew this piping strip around the edges of one panel, starting at the center of the back. Join the piping on the right side of the material using $\frac{1}{4}$in (6mm) seams, butting up the cord to make a neat finish.

Place the second panel on top of the piped cushion panel and, starting at one of the back corners, join together with $\frac{1}{2}$in (13mm) seams making sure that gathered corners come together. Leave the back edge open for insertion of the foam.

Straps and fixings

For the strap fastening, use curtain heading tape for contrast, or make up matching straps by covering the tape with the cushion material. Heading tape can be obtained in various colours and widths. For a 'different' strap, you could even try leather.

The straps are joined underneath, or behind, the furniture, using an ordinary buckle, or Velcro tape, or any other fastening that is convenient.

Where the straps cross, as with the seat cushions, fix them to the center of the cushion with a button threaded through the straps and cushion with nylon tufting twine. Fasten it underneath with another button (Fig. 13).

At any point where the cushion should be more permanently fixed to the frame to stop it moving, heavy-duty snap fasteners can be fixed to the frame and stitched to the fabric. A better alternative is Velcro tape, which can be stitched to the cusion and glued to the frame with a contact adhesive.

Parts List

Part	Description	Quantity	Length	Cut length to allow for waste	Size
CORNER CHAIR					
Legs	hardwood	3	$24\frac{7}{8}$ (627mm)	$25\frac{3}{8}$ (645mm)	1×4
		1	$11\frac{3}{4}$ (298mm)	$12\frac{1}{4}$ (311mm)	1×4 (25mm ×
Rails/arms		6	26 (660mm)	$26\frac{1}{2}$ (674mm)	1×4 102mm)
Seat supports		2	$24\frac{3}{4}$ (629mm)	$25\frac{1}{4}$ (642mm)	1×1 (25mm ×
		2	$22\frac{7}{8}$ (581mm)	$23\frac{3}{8}$ (594mm)	1×1 25mm)
Seat	plywood	1	$24\frac{3}{4} \times 24\frac{3}{4}$ (629mm × 629mm)	25×25 (635mm × 635mm)	$\frac{3}{8}$ (9mm)
Dowels	hardwood dowelling	18	$2\frac{1}{2}$ (64mm)	$2\frac{3}{4}$ (70mm)	$\frac{3}{8}$ (9mm) diameter
Bolts	GKN 2025	24	$1\frac{1}{2}$ (38mm)		$\frac{1}{4}$ (6mm)
Nuts	pronged 'T' Nuts	24			$\frac{1}{4}$ (6mm)
Screw-cups	M12 countersunk	24			No.12
Screws	Twinfast Pozidriv	12	$1\frac{1}{2}$ (38mm)		No.8
BACKED CHAIR					
Legs	hardwood	2	$24\frac{7}{8}$ (627mm)	$25\frac{3}{8}$ (645mm)	1×4
		2	$11\frac{3}{4}$ (298mm)	$12\frac{1}{4}$ (311mm)	1×4 (25mm ×
Rails/arms		5	26 (660mm)	$26\frac{1}{2}$ (674mm)	1×4 102mm)
Seat supports		2	$24\frac{3}{4}$ (629mm)	$25\frac{1}{4}$ (642mm)	1×1 (25mm ×
		2	$22\frac{7}{8}$ (581mm)	$22\frac{3}{8}$ (594mm)	1×1 25mm)
Seat	plywood	1	$24\frac{3}{4} \times 24\frac{3}{4}$ (629mm × 629mm)	25×25 (635mm × 635mm)	$\frac{3}{8}$ (9mm)
Dowels	hardwood dowelling	15	$2\frac{1}{2}$ (64mm)	$2\frac{3}{4}$ (70mm)	$\frac{3}{8}$ (9mm) diameter
Bolts	GKN 2025	20	$1\frac{1}{2}$ (38mm)		$\frac{1}{4}$ (6mm)
'T' Nuts	pronged 'T' Nuts	20			$\frac{1}{4}$ (6mm)
Screw-cups	M12 countersunk	20			No.12
Screws	Twinfast Pozidriv	12	$1\frac{1}{2}$ (38mm)		No.8
TABLE					
Legs	hardwood	4	$11\frac{3}{4}$ (298mm)	$12\frac{1}{4}$ (311mm)	1×4 (25mm ×
Rails		4	26 (660mm)	$26\frac{1}{2}$ (674mm)	1×4 102mm)
Seat/top supports		2	$24\frac{3}{4}$ (629mm)	$25\frac{1}{4}$ (642mm)	1×1 (25mm ×
		2	$22\frac{7}{8}$ (581mm)	$23\frac{3}{8}$ (594mm)	1×1 25mm)
Seat/top	plywood	1	$24\frac{3}{4} \times 24\frac{3}{4}$ (629mm × 629mm)	25×25 (635mm × 635mm)	$\frac{3}{8}$ (9mm)
Dowels	hardwood dowelling	12	$2\frac{1}{2}$ (64mm)	$2\frac{3}{4}$ (70mm)	$\frac{3}{8}$ (9mm) diameter
Bolts	GKN 2025	16	$1\frac{1}{2}$ (38mm)		$\frac{1}{4}$ (6mm)
Nuts	pronged 'T' Nuts	16			$\frac{1}{4}$ (6mm)
Screw-cups	M12 countersunk	16	$1\frac{1}{2}$ (38mm)		No.12
Screws	Twinfast Pozidriv	12	$1\frac{1}{2}$ (38mm)		No.8

Round table

There are only two simple types of joint used in the construction of this round table. The most complicated part is the top and even this really requires little more than patience to make, providing you use the correct tools. You will require three 5ft (1.5m) bar clamps to clamp the five table top planks together while the glue is setting, and four C clamps for the main frame joints. If you do not own bar clamps, it should not be too difficult to rent them locally.

Materials

The table shown here is made from clear pine, which was chosen for its attractive grain, but there are many other woods you could use. If you wish to paint the finished table you may use a cheaper grade of material. The dimensions of the parts are given in the cutting list.

The top is screwed to the frame underneath with No.8 steel screws. The joints should be glued with white glue, but only if the unit is to be used exclusively indoors. If it will be used in the open, or somewhere where it could get damp, a

Round table

waterproof adhesive such as a plastic resin or epoxy should be used.

For cutting out the circular table top, you will need an integral power jigsaw, or a jigsaw attachment for your power drill.

Construction details

The support structure – which in this case means all the members below the table top – consists of two interlocking frames as shown in Fig. 1. Each horizontal piece has an 'X' cross-lap joint cut out of the center of its narrow edge, flanked by two 'T' half-lap joints cut out of its wide face (Fig. 1D). The pairs of horizontal pieces intersect at the center 'X' halving; the additional cut-outs are to accept the vertical pieces for legs, which are halved at each end to form the 'T' halving joints.

The top is formed by gluing five pieces of lumber together to form a square approximately 48in (1219mm) each way, from which the circular top is then cut.

With regard to the top, when ordering these five pieces, ensure that the lumber is smooth and straight. If the pieces are not good and you have to plane them or compensate for their warp during assembly, the chances are that you'll end up with a poor job. Getting good boards to start with is a far easier course of action.

Making the top

The table top is made from wood $1\frac{3}{4}$in (45mm) thick and is a circle 48in (1219mm) in diameter when finished. You won't be able to purchase a plank as wide as this, so five 10in (254mm) wide planks should be butted together and glued along their long edges.

First arrange the planks side by side on a flat surface so that a 48in × 50in (1219 mm × 1270mm) (approx) rectangle is formed. The edges should butt against one another with no unsightly gaps in between; but if there are some high spots on the edges of the planks, carefully plane these down. Be very careful when doing this; set the blade very fine and make only a stroke or two at a time to avoid taking off too much wood. Otherwise, you may have to take off the whole edge to level it off again. If the high spots are only slight, buff them down with fine sandpaper wrapped around a block of wood.

Unless the top is to be painted, arrange these planks so that the grain forms an attractive pattern. Remember, this is the surface that will constantly be in view.

When you have finished fitting the planks together in a 'dry run' mark all the joints in sequence with one pencil line across the first joint and two across the second joint and so on, so that they can be lined up again in the correct order.

Next, glue the planks together, ensuring that all the joint marks line up exactly. Apply the three bar clamps using a small piece of wood at each clamp-shoe to prevent the clamp marking the lumber.

Apply the glue and clamp the pieces together. Spread adhesive evenly along the edges of the planks that are to butt, and clamp them together carefully. Place the first two clamps near each end of the planks with their bars running along and touching the underside of the planks. Place the third clamp near the middle of the planks with its bar running over the top surface of the planks. In this way, the planks are held flat on each side, which prevents them from jack-knifing when the clamps are tightened.

The planks must fit together perfectly to ensure that the top surface will be completely level. If one plank is slightly higher than the other, loosen the clamps a little, place a piece of waste wood on the raised plank and bang it with a mallet. This will force the raised plank down without marring the surface.

Re-tighten the clamps and wipe every trace of adhesive from the surface of the planks with a soft, damp cloth. If you don't do this thoroughly, any trace of adhesive will show through as a white patch when the wood is eventually varnished.

Allow adequate time for the adhesive to dry. 24 hours is usually enough, but follow the manufactuer's instructions. It is better to be overcautious than to spoil the job and have to do the work all over again.

The circle of the table top must now be marked out. Do this on the underside of the table top, using a nail, pencil and a length of string. Lightly tap a nail about $\frac{1}{2}$in (13mm) into the exact center of the panel. Tie a length of string to this, let the string out 24in (610mm) and tie the other end around a pencil as near the writing tip as possible. You can now swing the pencil around in a complete circle, marking the surface ready for cutting. For cutting the circle out of the glued planks, you will need a jigsaw. A thin piece of wood may be used in place of the string.

The important part is to ensure that

Figs 1A, B, C, D. The construction of the round table basic frame. The construction consists entirely of halving joints as shown in Fig 1F.

Fig 1E. The completed table top.

the top is held securely while you are running the jigsaw around the line. It must be clamped to a bench or table top with sufficient material protruding over the edge to allow you to make a part sweep around the line. Protect the surface with scrap wood blocks. When you have cut each sweep, loosen the clamps and move the top around to a position where you can continue to cut.

When you have cut the circle out, finish the edge with a plane, then sand it down.

Cutting the joints

Check your lumber dimensions when setting out the joints because the lumber you purchase may not be exactly $1\frac{3}{4}$in × $2\frac{1}{4}$in (45mm × 57mm). It will probably be slightly less, so mark out all joint recesses directly using the mating piece of lumber for accuracy.

Each horizontal member has an 'X' halving joint cut in the middle of its narrow edge. This should be as long as the member is wide, and exactly half the depth of the member, so that it will fit the mating horizontal piece tightly.

Next, cut the 'T' halving recesses for the legs. These must be the width of the member and as deep as half its thickness. One of these is cut on each side of the central cut-out with a space $2\frac{1}{4}$in (57mm) between each of the joints as shown in Fig. 1D.

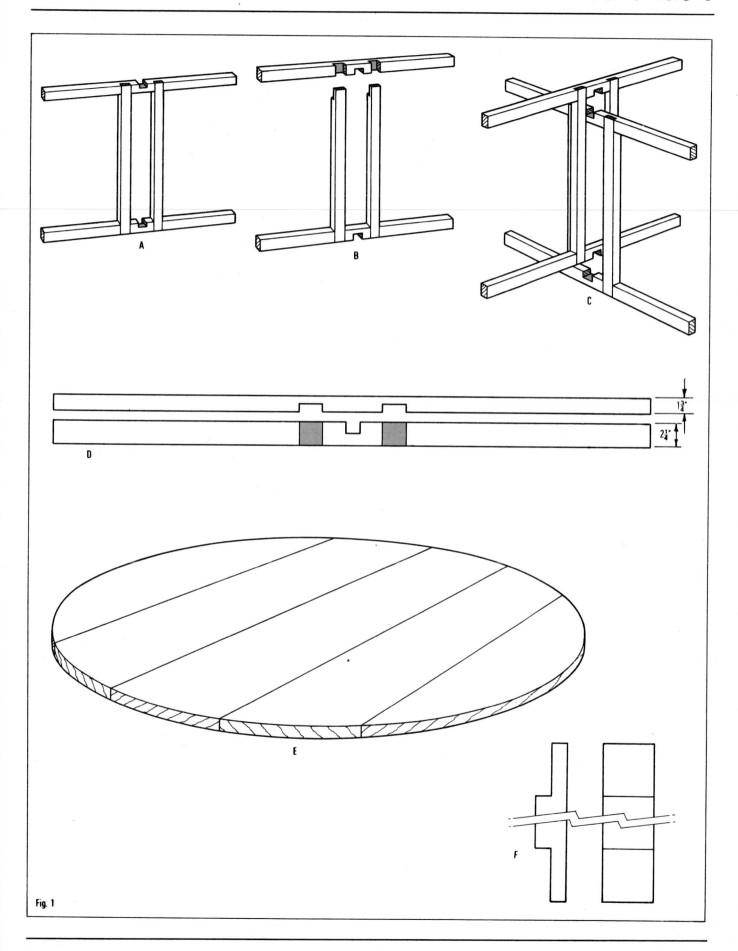

A

B

C

D

$1\frac{3}{4}"$

$2\frac{1}{4}"$

E

F

Fig. 1

Now cut the halving recesses in the ends of the legs following the same direct marking procedure and cutting to the same depth.

Using Fig. 1 as a guide, dry-assemble the two frames and interlock them. This will enable you to adjust any of the joints so that the surfaces are absolutely flush. When this has been done, with a pencil lightly mark the two mating parts of each joint with a number, so that you can re-assemble them in the same order.

Assembling the frames

The construction and fitting sequence for the frames is shown in Figs. 1A–C. Make up the first frame as in A. To do this properly, you need four C clamps, one at each halving joint. Spread adhesive on the surfaces of each halving joint and clamp the joints firmly in position, making sure that the assembly is square by measuring the inside diagonals of the rectangle formed in the center. Both diagonals should be the same length. If they are not, adjust the frame until they are. Wipe any excess glue from the joints and leave the frame for the adhesive to set.

When the adhesive has set, make up the second frame as shown in Figs. 1B and C.

First, fit the two legs into the halvings on the bottom rail. When these joints are set, put the assembly in position and glue the cross-halving joint between the two bottom rails.

Finally, fit the top rail in place gluing the rail halving and the leg joints at the same time. The base, or leg frame, is now complete.

Fitting the top

The table top is secured to the framework underneath with screws passing through the top horizontal members and into the underside of the top.

Drill countersink holes at 3in (76mm) intervals from edge to edge through the top horizontal members. The countersunk ends of the holes should be underneath the members. Place the table top on the framework, making sure that it is centered properly – the end of each top member of the leg framework should be the same distance away from the edge of the table top. Mark the position of pilot holes through the holes in the frame with an awl, remove the top, drill the pilot holes, and then replace the top and screw it firmly in position.

Finishing

If you are painting the table remember to use a good quality paint, particularly on the table top, because it will take a great deal of wear and this is one surface you don't want to look unsightly.

If you have chosen a natural wood finish, you will have to prepare the surface thoroughly so that it doesn't wear or mark easily, and at the same time bring out the rich grain of the wood.

Sand all surfaces with 100 grit sandpaper down to a smooth finish using an orbital finishing sander. Then use a soft cloth dampened with turpentine substitute to wipe all surfaces, to pick up the fine dust caused by sanding. With a good quality 2in (51mm) brush, sparingly apply a varnish of clear matt polyurethane and turpentine substitute diluted 50/50. When dry, rub down again with sandpaper and wipe the surface with the cloth. Now re-coat with the same varnish; when it is dry smooth it with grade 'O' steel wool and apply a final coat of undiluted polyurethane.

After a few days the unit can be polished with a soft cloth, and you will be the proud owner of a superb dining table.

Cutting list
Pine

	standard	metric
4 horizontal members	$2\frac{1}{4} \times 1\frac{3}{4} \times 43$	$57 \times 45 \times 1092$
4 vertical members	$2\frac{1}{4} \times 1\frac{3}{4} \times 27$	$57 \times 45 \times 686$
5 table top planks	$1\frac{3}{4} \times 10 \times 48$	$45 \times 254 \times 1219$

You will also require:

16 $3\frac{1}{2}$in (93mm) No.8 flathead wood screws. Adhesive. Varnish and turpentine substitute or paint.

For this project you will need 5ft (1524mm) bar clamps.

The sizes for the members are the finished dimensions and do not include any allowance for waste. The sizes for the table top planks are the dimensions before the pieces have been glued together and the circle cut out.

Table with a rural look

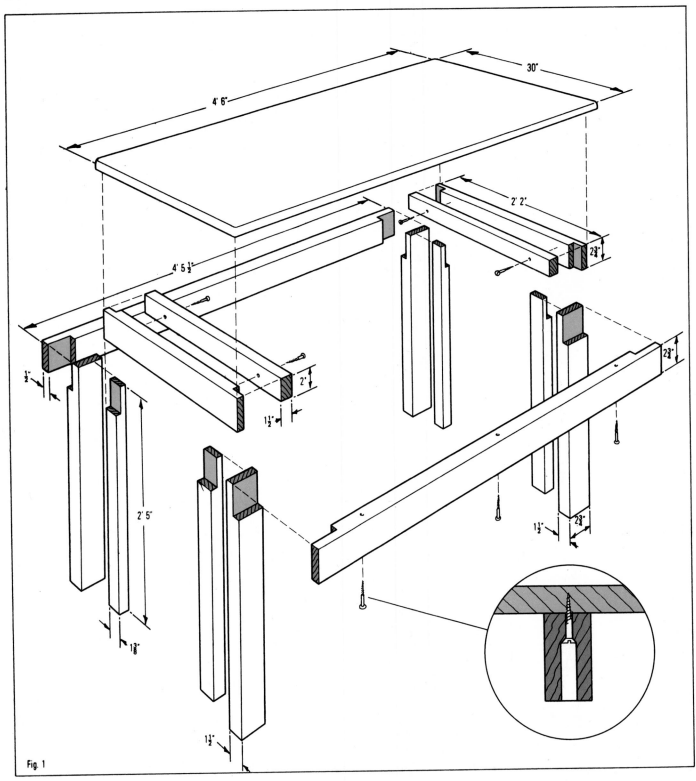

Fig. 1

Fig 1. Exploded view of the rural table. All leg to rail joints are half laps. Each side is constructed as a separate unit and joined at a later stage.

This solidly built furniture will complement any home. The strong grain pattern of the tops give a hint of old fashioned country charm, blending well with the more modern finish on the legs. The simple construction of the table and benches makes them an attractive pro-proposition for the less experienced

woodworker. The only jointing involved is the cutting of the lap joints.

Materials

The dimensions of the finished components of the table and benches are given in the cutting list. The wood used in the units illustrated is pine, but any softwood with an attractive grain and capable of being sanded to a smooth finish will do.

You will also need some No. 8 flathead steel screws and some glue. If you intend to use the furniture solely indoors, a carpenter's glue is ideal, but a waterproof glue such as Resorcinol should be used if the table and benches are likely to be used outdoors.

Construction

The main feature of these units is a compound leg, stronger than the usual type of kitchen table leg but lighter and hence more attractive looking. In these instructions, the larger component is described as the 'outer leg' and the smaller as the 'inner leg'.

The two rails that run parallel to the long sides of the table and bench tops are half-lap jointed to the outer legs. The rails that run parallel to the short sides are half-lap jointed to the inner legs, and this U-shaped construction is fixed in place between the long rails and the outer legs. Fig. 6 shows a cross-section of the finished legs.

The rails and legs are screwed to the underside of the benches and table. On the table only, extra support for the top is provided by two bearers which run parallel to the two short rails and are butted against them.

The cross-sections of the wood in the legs and rails of the table are exactly the same as those used for the legs and rails of the benches.

Making the top

The top of the table is made from wood $\frac{3}{4}$in (19mm) thick and has a finished width of 30in (762mm). You will not be able to buy a single plank of this width so three planks, each 10in (254mm) wide, are butted and glued along their long edges. You will find that 10in (254mm) is not a standard width of wood dressed on four sides (D4s); so-called 10in (254mm) board starts out 10in (254mm) wide when rough sawn, but is reduced in the planing process. So you will have to buy wider boards and, for reasons which follow, it might be wise to buy three 12in (305mm) planks for the table top and one 14in (356mm) for each bench top.

The finished width of the bench tops is $11\frac{3}{4}$in (299mm). You should be able to obtain single planks of this width, but as 12in (305mm) D4 lumber is now only $11\frac{1}{4}$in (287mm), you will have to buy

$\frac{3}{4}$in × 14in (19mm × 356mm).

So select your pieces carefully at the lumberyard. If you have difficulty you can butt two planks of half this width along their long edges, but this obviously takes longer and does not give so neat a finish.

The first step in making the table top is to cut the planks slightly longer than the finished length – about 1in (25mm) too long. Then lay the planks together with their long edges touching. Since the top is later finished in clear polyurethane, choose the face of each plank which has the most attractive grain and lay them out to make the best pattern as a whole. Do this on a dead-flat surface to make sure that any cupping or bowing in the pieces is in the same direction; otherwise you will have difficulty obtaining a smooth top surface. Then mark the faces lightly with a pencil (fig. 3) to ensure that the right edges are butted and that you do not accidentally reverse one of the planks.

Now plane the edges that are to butt exactly square. Use a long plane (jack or jointer) to do this – a short smoothing plane will simply follow the contours of the wood (Fig. 2). Check that the planed edges are perfectly flat and level by drawing a straightedge along their surfaces.

The next step is to glue the planks together, using either a carpenter's glue or a waterproof glue. To do this you will need three bar clamps and some waste blocks of wood. So that you can quickly join the planks once the adhesive has been applied, pre-set the clamps to the correct length by clamping the planks together dry. Place the waste blocks between the shoes of the clamps and the edges of the planks. This will spread the clamping pressure more evenly and protect the edges of the planks from damage.

Now glue can be applied and the pieces clamped together. To do this, first release the clamps. Spread adhesive evenly on the edges of the planks that are to butt. Clamp them together again. Position one clamp near each end of the plank with their bars running along the undersurface of the planks. Place the third clamp near the middle of the planks with its bar running across the top surface of the planks.

The planks must fit together perfectly with their surfaces completely level. If one plank is slightly higher than the others, loosen the clamps a little. Place a waste piece of wood on the raised plank and

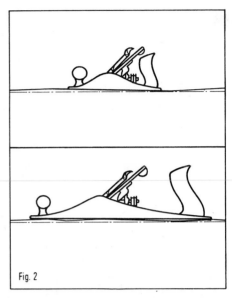

Fig. 2

Fig 2. When planing the planks for the top a jack plane will give better results.

bang it with a hammer. This will force the raised plank downwards so that its surface is flush with the other planks.

Now re-tighten the clamps. Wipe every trace of adhesive from the faces of the planks, using a soft damp cloth. Any adhesive left on the surface will show as a white patch under the polyurethane finish that is applied later.

Check that there are no irregularities in the planks by running your finger along the joints. Allow adequate time for the adhesive to dry. A day is usually long enough, but follow the manufacturer's instructions. It is better to be over-cautious than spoil the job and have to do the work all over again.

When the adhesive has dried the top can be cut exactly to length and width. Square a line near the end of the planks right across the top face. The finished length of the table top is 4ft 6in (1.37m) and the finished dimensions of the bench tops will be $11\frac{3}{4}$in × 3ft $11\frac{1}{2}$in (298mm × 1.20m). Measure these distances on the relevant pieces from the squared line. Square lines through these new marks.

The next steps are to saw just outside the marked lines, then plane away the waste. You must be careful, however, not to damage the corners of the planks, which is always a danger when planing end grain. There are three ways of doing this.

The first method is to clamp a waste block of wood to the end which you will be planing. The top of the waste block should be level with the top edges of the

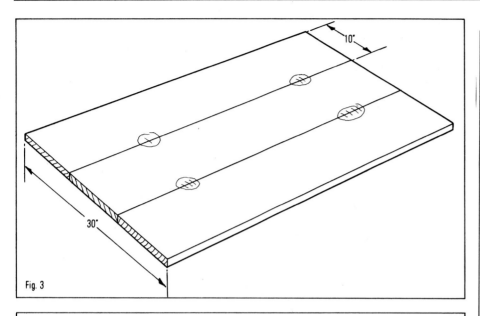

Fig. 3

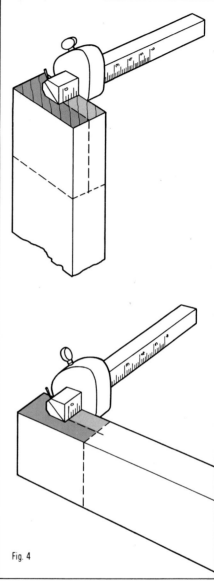

Fig. 4

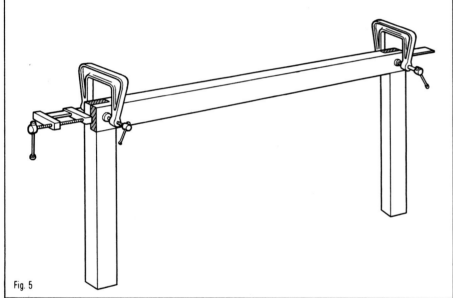

Fig. 5

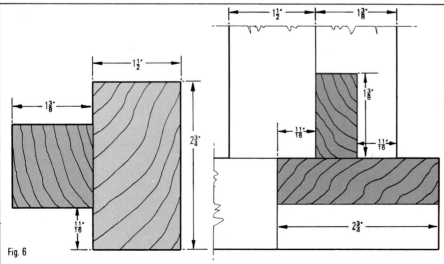

Fig. 6

Fig 3. The planks should be checked for fit and marked to make sure the final fit is tight.

Fig 4. The method used to mark the half lap joints for the legs and cross rails.

Fig 5. The legs and cross rails are clamped together using two C clamps and one bar clamp.

Fig 6. The way in which the legs and cross rails join together.

planks so that any damage that is done when planing will be to the waste block. The second method involves planing from each end of the planks toward their center. The third method is to chisel away the waste at one corner for up to 1in (25mm) and then plane from the opposite corner. But you must be careful, when chiselling, not to cut away any wood below the squared line.

Using one of these methods, finish the top to the required size with the ends perfectly square.

Jointing the rails and legs

The thickness of the legs and rails of the table is the same as that of the rails and legs of the benches. The method described below of making one joint therefore, applies to all the joints.

Joint the outer legs and rails first. Cut the rails 1in (25mm) oversize and mark out the positions of the joints on both pieces at the same time. Place the rails together face-side to face-side, then square a line across one end and, from that line, measure the length of the rail and square a line across at the other end of the lumber. From these lines set out the width of the legs, using the prepared timber as a pattern. Separate the rails, mark the face sides and square the lines around all sides.

Set a marking gauge to exactly half the thickness of the lumber and working from the face-side score a line on the edge of the rail and across the end grain (Fig. 4) to indicate the depth of the halved joint.

Set out the legs in the same way. First, cut them overlength and mark the length. Use the rails to set out the width of the joint. Score the depth of the joint with the marking gauge working from the face-side as before. Shade the inside face of the rail joints with a pencil so that you will know this is to be cut out. Then shade or mark with a cross the outside of the leg joint as this will be the waste side.

Cut down the waste side of the line using a tenon saw. Then, score over the pencil joint-line, across the grain, using a handyman's knife and a try-square. Saw down this line to remove the waste. The scored line will give a neater finish to the shoulders of the joint.

When all the half laps in the long rails have been cut and tried for fit, the rails can be sawn to only $\frac{1}{16}$in (1.5mm) longer than their finished length at each end. This will be smoothed off later.

The half lap at the top of the legs can

now be cut. The method is exactly the same as that for the rails. As each leg is cut, lightly mark with a pencil both the leg and the rail into which it fits, so that the components do not get mixed up during final assembly.

The short rails and inner legs

Now cut the short rails and the inner legs. The rails are the same width and thickness as the long rails – 1½in × 2¾in (38mm × 70mm) – and are a finished length of 2ft 2in (660mm) for the table legs and 8¼in (209mm) for the bench legs. The legs are 1⅜in × 1½in × 2ft 5in (34mm × 38mm × 737mm). These again are jointed with lap joints which are marked out and cut in the manner described above with, of course, the necessary adjustment for the different dimensions. Note: The dimension of the leg length is the finished dimension. At this stage in the construction, you should be sure to allow 1in (25mm) waste for levelling.

Fixing the legs to the rails

Now the rails and legs can be glued together. Apply adhesive to both faces of the lap joints and push the pieces together. Lightly clamp the two laps at each end together with a C clamp. The C shaped bar should run over the top edge of the rail. Then apply a bar clamp so that the bar of the clamp runs along and parallel to one face of the rail (Fig. 5). Check with a try-square that the angles between the legs and rails are perfectly square and tighten all the clamps. Wipe any excess adhesive from the surfaces of the pieces and allow adequate time for the glue to dry.

When the glue has dried, use a sharp smoothing plane to remove the small pieces of the lap joints that protrude.

Fixing the frame together

When the legs and rails of the table and benches are of the finished size, the inner legs can be fixed inside the long rails and the outer legs. The outside face of the inner legs is butted to the inside face of the outer legs. It is essential to work accurately here or the frame will not be square.

First mark out the correct positions of the inner legs on the insides of the outer legs. The center of the edge of the inner legs should run along the exact center of the inside face of the outer legs. In order to find the correct position, first mark a line along the center of the inside face of the outer legs. Then mark on each side of this line half the thickness of the inner leg – which is 1⅜in (35mm). Mark a line through this point parallel with the long edges of the legs.

Fix the short rail and leg assembly and the long rail and leg assembly together with adhesive. To ensure that the pieces do not move out of position while the glue is wet, knock finishing nails partway into the inside face of the outer leg along the two lines that indicate the correct position of the inner legs. Space the nails about 2in (50mm) apart and knock them in at an angle as shown in Fig. 7. This will enable you to remove them more easily when the adhesive has dried.

Now glue the pieces together. Apply adhesive to the face of the inner legs and

Fig 7. Nails are used to align the inside legs.

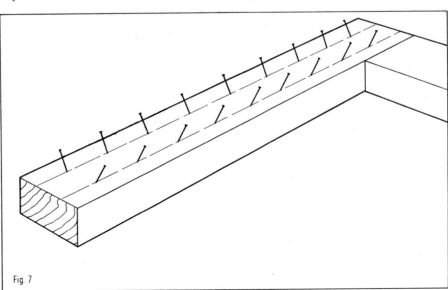

Fig. 7

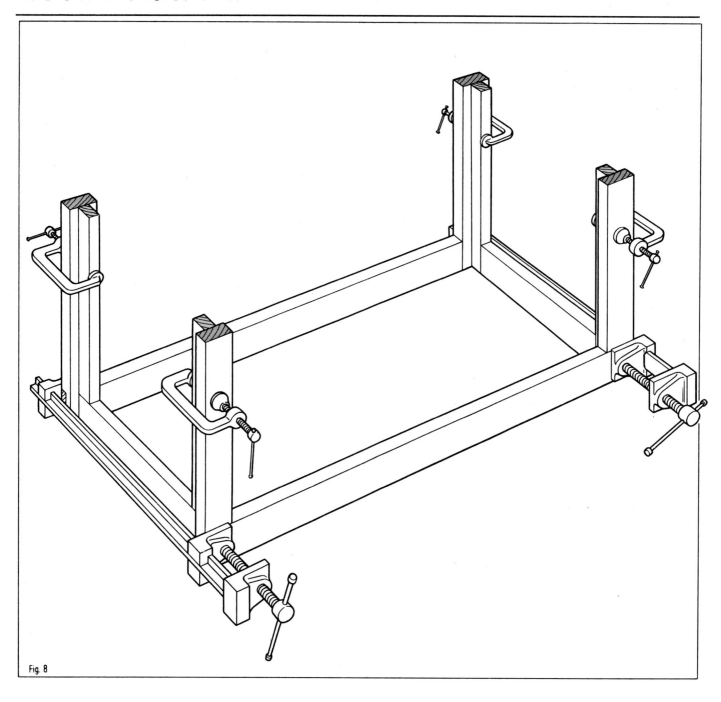

Fig. 8

Fig 8. The finished leg and rail assemblies are glued and clamped together using bar and C clamps.

push the four components of the frame together. Before clamping them, turn the assembly upside down on a perfectly flat surface so that the legs stick up in the air. Use bar clamps along the level of, and parallel to, the rail of the inner leg assembly. Apply only light pressure at first. Then check that the frame is perfectly square and tighten the bar clamps. Use C clamps to clamp the outer legs to the inner legs (Fig. 8). Wipe any excess adhesive from the surfaces of the legs. Allow the adhesive to dry. When the adhesive has dried remove the finishing nails. The holes can be filled with wood

filler, but they are not in a conspicuous place.

Finishing the furniture
The frame and legs and the top are finished before the assembly is complete. Sand down all surfaces. Apply a clear polyurethane to the tops of the tables and benches. Use several coats, rubbing down the previous coat with steel wool before applying the subsequent coats. The varnish should be applied to both the top surface and the undersurface of the tops.

The finish applied to the legs is a matter of taste. The entire unit can have a natural

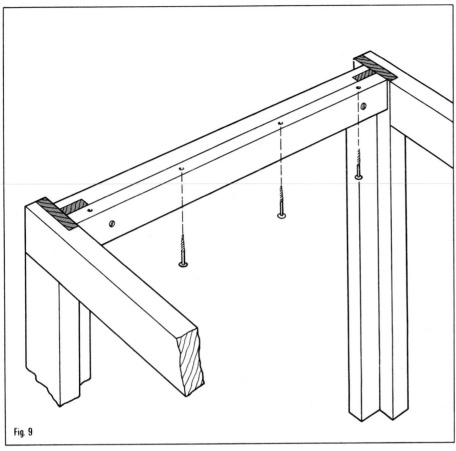

Fig. 9

Fig 9. The table top is screwed to the table via a bearer which fits inside the ends of the table. If the color of the top is to be different to that of the frame it is advisable to finish both components before screwing together.

Fig 10. Exploded view of the bench. The method of construction is exactly the same as that of the table.

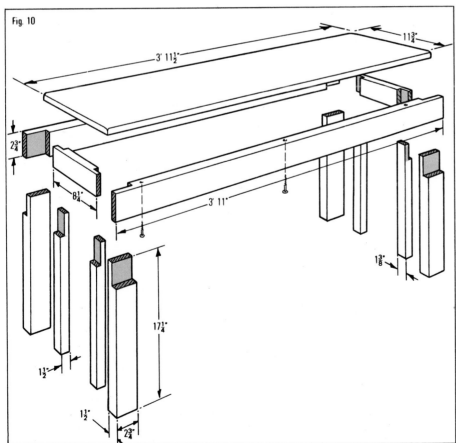

Fig. 10

finish, like the top; or you can choose a contrasting color such as the polyurethane finish in the photograph.

Fixing the frame to the top

The framework of the tables and benches is screwed to the tops. They cannot be glued in place first as this will spoil the polyurethane finish – and the two-color finish cannot be applied neatly once the pieces have been fixed.

The frame must be held in its correct position while it is being screwed to the top. This can be done by laying a plank across the bottom edge of the rails and kneeling on it while you fix two screws.

In the case of the benches these first screws can be fixed in the center of each short rail. For the table, however, the first two screws should be fixed through the long rails, one at each end and diagonally opposite. This is because holes drilled through the bottom edge of the short rails of the table will tend to be conspicuous (this is not, however, the case with the benches). The inner leg assembly of the table is fixed to the top by means of bearers which are screwed to the table top and to the inside face of the short rails, creating a neater finished appearance.

The frame is fixed to the top with $1\frac{1}{2}$in (38mm) No. 8 flathead steel screws. The heads of the screws are buried in the rail through a second hole the diameter of the screw head, which should be drilled part of the way into the rail down the first hole. This will do away with the need for unnecessarily long screws on the job.

When drilling through the rails into the top you must be very careful not to cut the holes too deeply. The rails are $2\frac{3}{4}$in (70mm) wide and the tops of both the tables and the benches are $\frac{3}{4}$in (19mm) thick. The first drilled holes, therefore, should be about 3in (76mm) deep. To cut these holes to an accurate depth, wrap a piece of colored adhesive tape around the drill this distance from the drill tip. You can then stop drilling as soon as you reach the tape. The same method can be used to drill the countersinking holes, which are the diameter of the screw heads. These should be drilled to a depth of $1\frac{3}{4}$in (44mm).

Mark the positions of the holes on the rails. For the benches use one screw in the center of each short rail and three along the long rails. Position one centrally and one near each end. For the table use three screws in each long rail, positioned as for the benches. Do not drill any holes in the short rails of the table.

Table top bearers

The next step in the construction of the furniture is to fix bearers to the table top. These measure $1\frac{1}{2}$in $\times$ 2in $\times$ 2ft $1\frac{3}{4}$in (38 mm $\times$ 50mm $\times$ 654mm) and fixed so that one face butts against the inside face of the short rails of the inner leg assemblies. Use two 2in (50mm) No. 9 countersunk steel screws to fix the bearers and ensure that you do not drill too deeply by using colored tape around the drill, as described previously. To fix the bearers to the top, use four 2in (50mm) No. 8 flathead steel screws for each bearer. Position one screw about 2in (50mm) from each end of each bearer and one screw about 7in (180mm) from the first, toward the center of the bearer.

Cutting the legs to length

This is the final stage in the construction of the table and benches. The legs were originally cut $\frac{1}{2}$in (12mm) too long. You do not have to remove exactly $\frac{1}{2}$in (12mm), provided you cut the right amount of wood from all the legs of each article. It is also likely that at least one of the pieces will wobble when placed on a flat surface. This is because of unavoidable inaccuracies occurring while working.

To cut the legs to an equal length stand the bench on a flat surface. If there is any wobble, caused by one leg being shorter than the other, place waste pieces of wood under the short leg until the piece stands level.

Then take a waste block of wood measuring about 1in $\times$ 2in $\times$ 6in (25mm $\times$ 50mm $\times$ 152mm). Drill a hole slightly less than the diameter of a pencil, through the 1in (25mm) edge. With a mallet knock the pencil through the hole so that about 1in (25mm) of it protrudes. Then lay the waste block on the flat surface so that the pencil is parallel with the floor and its point is touching the leg. Draw a line around the bottom of each leg. Now carefully cut through the legs, around these lines.

Sand the sawn ends smooth and, finally, touch up any damage to the ends of the legs with polyurethane paint or varnish, depending on the finish you have chosen.

Cutting lists
Table
Solid wood

	standard	metric
3 tops	$\frac{3}{4}$in $\times$ 10in $\times$ 4ft 6in	19mm $\times$ 254mm $\times$ 1.37m
2 long rails	$1\frac{1}{2}$in $\times$ $2\frac{3}{4}$in $\times$ 4ft $5\frac{1}{2}$in	38mm $\times$ 70mm $\times$ 1.36m
2 short rails	$1\frac{1}{2}$in $\times$ $2\frac{3}{4}$in $\times$ 2ft 2in	38mm $\times$ 70mm $\times$ 660mm
4 outer legs	$1\frac{1}{2}$in $\times$ $2\frac{3}{4}$in $\times$ 2ft 5in	38mm $\times$ 70mm $\times$ 737mm
4 inner legs	$1\frac{3}{8}$in $\times$ $1\frac{1}{2}$in $\times$ 2ft 5in	34mm $\times$ 38mm $\times$ 737mm
2 top bearers	$1\frac{1}{2}$in $\times$ 2in $\times$ 2ft $1\frac{3}{4}$in	38mm $\times$ 50mm $\times$ 654mm

For each bench
Solid wood

1 top	$\frac{3}{4}$in $\times$ $11\frac{3}{4}$in $\times$ 3ft $11\frac{1}{2}$in	19mm $\times$ 299mm $\times$ 1.20m
2 long rails	$1\frac{1}{2}$in $\times$ $2\frac{3}{4}$in $\times$ 3ft 11in	38mm $\times$ 70mm $\times$ 1.19m
2 short rails	$1\frac{1}{2}$in $\times$ $2\frac{3}{4}$in $\times$ $8\frac{1}{4}$in	38mm $\times$ 70mm $\times$ 209mm
4 outer legs	$1\frac{1}{2}$in $\times$ $2\frac{3}{4}$in $\times$ $17\frac{1}{4}$in	38mm $\times$ 70mm $\times$ 438mm
4 inner legs	$1\frac{3}{8}$in $\times$ $1\frac{1}{2}$in $\times$ $17\frac{1}{4}$in	35mm $\times$ 38mm $\times$ 438mm

These sizes are to the finished dimensions of the table and benches and do not include any allowances for waste.

Dining chairs

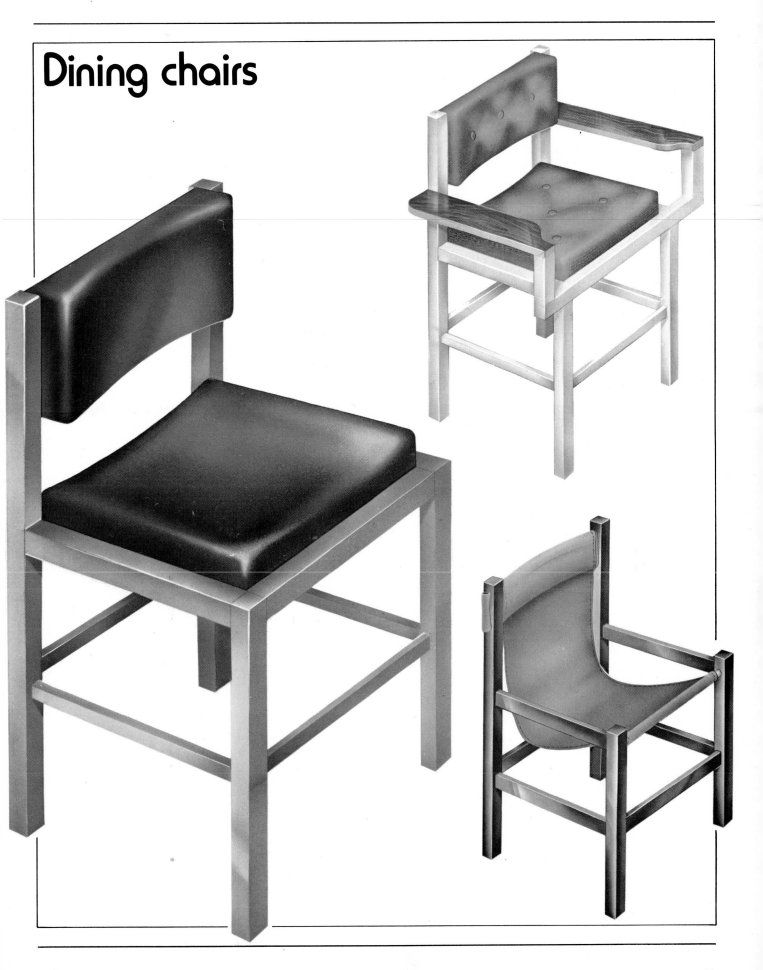

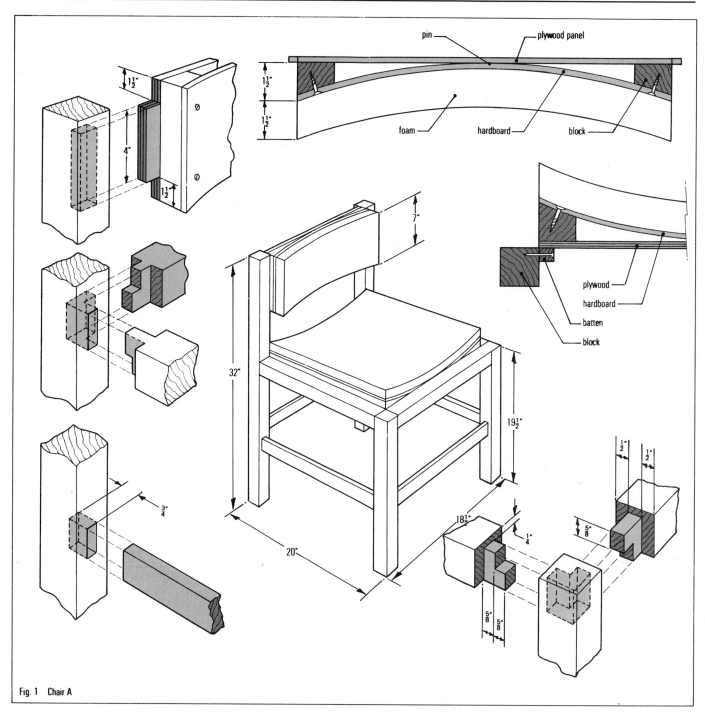

Fig. 1   Chair A

Fig 1. Details of the construction of Chair A —suitable for use as either a dining room or kitchen chair.

Of the four chairs described, the first three are all built around one design. This design has been carefully worked out with four aims in mind: first, to be attractive; second, to be of sturdy construction; third, to be simple to sonstruct and last, to be adaptable enough to allow individual modifications.

These aims have been met. All the chairs are constructed using simple but strong joints and they are devoid of unnecessary details. In this respect they are ideally suitable for the home carpenter

who wishes to build a matching set of chairs without an overwhelming amount of work and expense. For simplicity the chairs are called A, B, C and D.

Chair A

Chair A (illustrated in Fig. 1) is suitable for use both as a dining chair and a kitchen chair. You have a wide choice of wood, and this can largely be dictated by the intended use of the chair. If, for example, it is to be a dining chair, a hardwood, such as beech, elm, teak or mahogany gives a pleasantly sophisticated appearance. On the other hand, if you intend using the chair in the kitchen, a softwood, such as pine (painted or naturally finished) presents a more casual appearance.

Cutting out

Cut out all the pieces to the sizes given in the cutting list, taking particular care to ensure that the legs are all the correct length. You will find it easier to sand all the separate members at this stage, rather than when the chair is assembled. Use coarse sandpaper, then fine, and take off the sharp edges but do not round them.

The joints

You can follow one of two procedures to cut out the joints. You can either cut out all the mortises at once, then all the tenons, or you can work in strict order and make each matching mortise and tenon before going on to the next. The second method is more suitable for the home handyman as it allows you to check each joint for fit as it is cut.

Start by putting all four legs together with the bottom ends exactly level. Then square lines across them to mark the positions of the upper and lower rails. This will ensure that all the mortises are in line with each other. You can either set out all the mortises together or one at a time using the squared lines for getting the correct position.

The lower rails do not have tenons cut in them, they simply slot straight into the mortised holes. The lower rails are 8½in (216mm) from the bottom of the leg to the bottom edge of the rail. Cut the mortises to a depth of ¾in (19mm), and when the waste wood has been removed, trial assemble the lower rails to the legs and, if necessary, trim for fit.

Now cut the joints for the upper side rails. These are more complicated but, if care is taken, should present no real problems. Begin by cutting out the mortises for the side rails in the front legs. These mortises are 1¼in (32mm) deep, 1in (25mm) long, ½in (13mm) wide and are located ¼in (6mm) below the top edges of the front legs. Then cut the corresponding tenon joints on the front ends of the side rails. In this case these joints are haunched tenons, which are cut in such a way that they lock into the corresponding joints cut into the upper front rail.

To cut the haunched tenons, first cut

out a conventional tenon ½in (13mm) thick and 1in (25mm) wide. Then, using Fig. 1 as a guide, cut a ¼in (6mm) strip off the top of the tenon. Next, form the haunch by cutting a piece ⅝in (16mm) square out of the top edge of the tenon.

When the side rails have been completed, cut the mortises for the front and back rails. The tenons are the same as for the front rails, but the haunch is reversed so that is will interlock with the side rail tenons. Fit all the joints dry to ensure thet they are tight and that the rails are parallel.

The backrest

First cut out mortises 4in (102mm) long, ¾in (19mm) deep and ¼in (6mm) wide in the rear legs. These mortises are located so that their top edges are 1½in (38mm) from the top ends of the rear legs. Make corresponding tongues on the plywood back, as shown in Fig. 1. Take the two wood blocks and cut them out to the shape shown in Fig. 1. Glue and screw the blocks to the plywood back, then glue and screw the hardboard back to the blocks. This hardboard back is then held in a curved position by nailing it, at the center, to the plywood back panel.

The seat

This part is assembled in the same way as the backrest. A ¼in (6mm) plywood panel is situated on battens which are glued and screwed to the top rails (Fig. 1), and to this panel are fixed full-length shaped blocks and a curved hardboard seat panel.

Final assembly

With all the joints cut, assemble the chair in the following order.

Glue one of the lower side rails into the front and back legs. Before the glue sets, add the upper side rail. Lay this assembly on its side and fix the upper front and back rails into position, followed by the lower front and back rails.

Leave this assembly while you add the other side rails and legs. Then fix them to the chair assembly and add the backrest. Allow the glue to set and, if possible, use clamps to hold each member in its correct position. When the glue has set add the seat assembly.

Covering the chairs

Both the backrest and the seat are covered with 1½in (38mm) thick foam rubber. You can choose from a wide range of

materials for final covering, including both fabrics and leather.

Depending on what wood you have used, you can give the chair a varnish, paint or natural finish.

Chair B

Popularly called a carver chair, the design of this chair differs in that it includes a pair of armrests. There are some other modifications, including a longer upper front rail and shorter front legs.

To make it, first cut out all the pieces to the sizes given in the cutting list. The rear of the chair (including the backrest), the lower rails and the seat are assembled in exactly the same way as described for chair A. Where the front construction differs is in the assembly of the front legs to the top front rail, and the addition of the armrests.

The front legs are recessed into the top front rail and are located so that their outside edges are 3in (76mm) from the ends of the front rail. The joints used are mortise and tenon types and these lock into corresponding joints cut in the upper side rails (Fig. 2). The armrests are supported at the front on two members which are through-mortise and tenoned into the front rail, as shown in Fig. 2. Mortise and tenon joints are also used to join the armrests to the supports.

A different jointing technique is employed to attach the armrests to the rear legs. The holes are bored in the narrow edges of the armrests where they coincide with the rear legs. Matching holes are then bored in the rear legs, and short lengths of dowelling glued into the matching holes to form a blind dowelled joint.

Final assembly follows the same order as that employed for chair A, and the chair can be finished in the same ways.

Chair C

This sling-back chair uses the same basic framework as the previous chairs, but its appearance is transformed by the substitution of a canvas slung seat for the rigid wooden seat. Apart from this feature, the only other modification is that the upper side and back rails have the same dimensions as the lower rails and are jointed into the legs in the same way. This makes the assembly of this chair much easier, as there are no complicated haunched mortise and tenons to cut.

Begin by cutting out all the pieces to the sizes given in the cutting list. Make

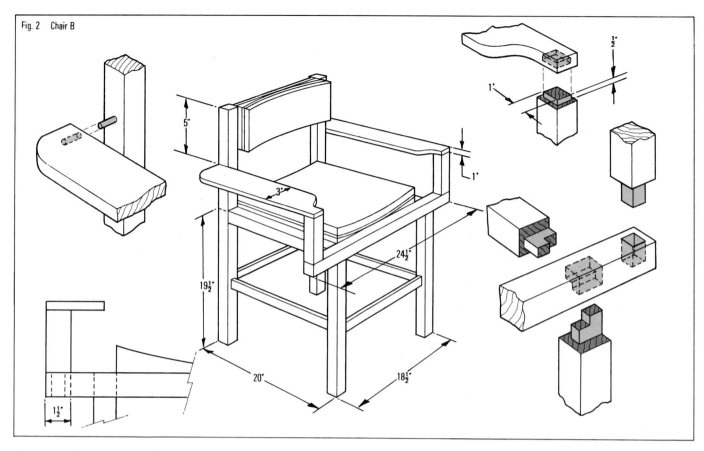

Fig 2. *Details of the construction of Chair B. Fig 3. Chair C and its supporting leather strap.*

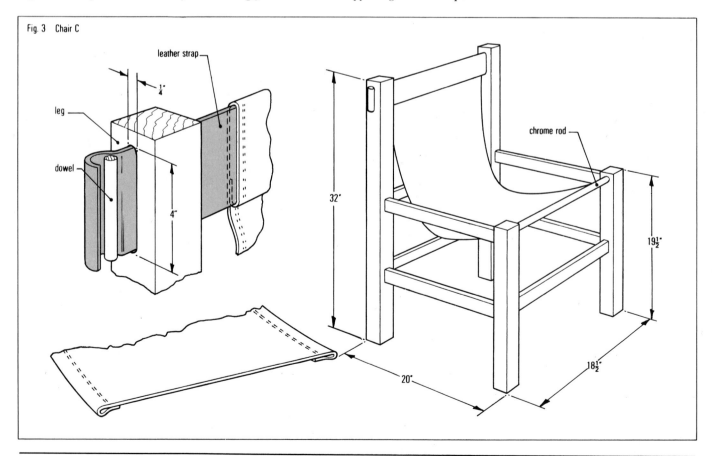

the joints as detailed previously. In place of the upper front rail is a $\frac{3}{4}$in (19mm) diameter chrome tube which is glued into holes bored into the front legs, as shown in Fig. 3. Then cut out slots (or through-mortises) in the back legs to take the hide strap which acts as a backrest. These slots, which are illustrated in Fig. 3, are located $\frac{3}{4}$in (19mm) below the top ends of the back legs, and should be just wide and long enough to take the hide strap.

When the framework of the chair has been assembled, place the backrest in position. This is done by first passing the hide strap through the slots so that an equal length of strap protrudes from both slots. Then, take lengths of dowel just longer and wider than the slots, and sew or stud the strap around the dowel and back on itself, as shown in Fig. 3.

The slung seat can be made from a strong fabric, such as heavy canvas, or from leather or some synthetic material. Before fitting it, turn it back and sew the long edges, as shown in Fig. 3, to prevent it from tearing under the weight of a person. Attach it first to the hide strap, as shown in Fig. 3, then gauge the amount of slack needed before fixing the other end around the chrome tube.

Depending on what material you choose for the seat, this chair can be used for a variety of purposes. Fitted with a leather seat it makes an elegant and

Cutting list: Chair A

Solid wood	standard	metric
2 back legs	$1\frac{1}{2} \times 1\frac{1}{2} \times 32$	$38 \times 38 \times 813$
2 front legs	$1\frac{1}{2} \times 1\frac{1}{2} \times 19\frac{1}{2}$	$38 \times 38 \times 495$
2 lower side rails	$\frac{1}{2} \times 1 \times 18\frac{1}{2}$	$13 \times 25 \times 470$
2 upper side rails	$1\frac{1}{2} \times 1\frac{1}{2} \times 18\frac{1}{2}$	$38 \times 38 \times 470$
2 lower cross rails	$\frac{1}{2} \times 1 \times 17$	$13 \times 25 \times 432$
2 upper cross rails	$1\frac{1}{2} \times 1\frac{1}{2} \times 17$	$38 \times 38 \times 432$
4 battens	$\frac{1}{2} \times \frac{1}{2} \times 15$	$13 \times 13 \times 381$
2 wood blocks	$1\frac{1}{2} \times 1\frac{1}{2} \times 7$	$38 \times 38 \times 178$
4 wood blocks	$1\frac{1}{2} \times 1\frac{1}{2} \times 17$	$38 \times 38 \times 432$
Plywood		
1 back panel	$\frac{1}{4} \times 7 \times 17$	$6 \times 178 \times 432$
1 seat panel	$\frac{1}{4} \times 15\frac{1}{2} \times 17$	$6 \times 394 \times 432$
Hardboard		
1 seat panel	$\frac{1}{4} \times 15\frac{1}{2} \times 17$	$6 \times 394 \times 432$
1 back panel	$\frac{1}{8} \times 7 \times 15\frac{1}{2}$	$3 \times 178 \times 394$

You will also require:
Foam rubber padding. Carpenter's glue. $\frac{1}{2}$in (13mm) finishing nails.

Cutting list: Chair B

Solid wood	standard	metric
2 back legs	$1\frac{1}{2} \times 1\frac{1}{2} \times 32$	$38 \times 38 \times 813$
2 front legs	$1\frac{1}{2} \times 1\frac{1}{2} \times 18$	$38 \times 38 \times 458$
1 upper front rail	$1\frac{1}{2} \times 1\frac{1}{2} \times 24\frac{1}{2}$	$38 \times 38 \times 622$
2 armrests	$1 \times 4 \times 20$	$25 \times 102 \times 508$
2 armrest supports	$1\frac{1}{2} \times 1\frac{1}{2} \times 7\frac{1}{2}$	$38 \times 38 \times 191$

All other members are the same dimensions as the corresponding members of Chair A.

Cutting list: Chair C

Solid wood	standard	metric
2 back legs	$1\frac{1}{2} \times 1\frac{1}{2} \times 32$	$38 \times 38 \times 813$
2 front legs	$1\frac{1}{2} \times 1\frac{1}{2} \times 21$	$38 \times 38 \times 534$
4 side rails	$\frac{1}{2} \times 1\frac{1}{2} \times 18\frac{1}{2}$	$13 \times 38 \times 470$
3 cross rails	$\frac{1}{2} \times 1\frac{1}{2} \times 17$	$13 \times 38 \times 432$

You will also require:
1 $\frac{3}{4}$in (19mm) diameter chrome tube 17in (432mm) long. Hide strap $22\frac{1}{2}$in $\times$ 4in $\times \frac{1}{4}$in (572mm $\times$ 102mm $\times$ 6mm). Leather or fabric 36in $\times$ 19in (915mm $\times$ 482mm).

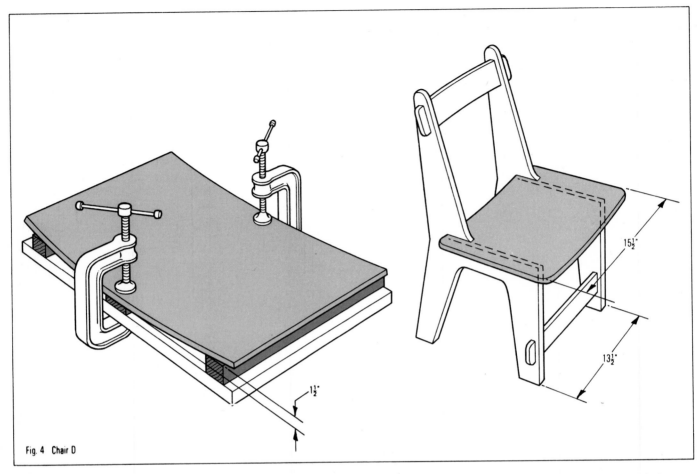

Fig. 4 Chair D

Fig 4. Construction details of Chair D. Also shown is the method used to curve the plywood for the back and seat. It is always a good idea to use waste blocks between the shoes of the clamps and the work to prevent bruising of the wood.

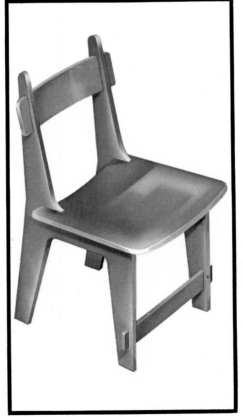

unusual dining chair. If a colorful canvas seat is chosen, it looks equally good in the kichen or in the garden.

Chair D

If you want to make a matching set of chairs easily and cheaply, this design is ideal. The material used is $\frac{5}{8}$in (16mm) birch plywood, the sides being cut from one sheet of ply 25in × 32in (635mm × 813mm), while the seat, backrest and rail are cut from one sheet $15\frac{1}{2}$in × 21in (394mm × 533mm). (Note that an allowance must be made for the thickness of the saw cuts. The seat and backrest are slightly curved to fit the contours of the body. To curve plywood, first apply a heavy coating of water-based wood glue to the wood, then fix it in the desired shape with a C clamp, as shown in Fig. 4. Leave in a warm place for 48 hours by which time the glue will have set and will hold the curve. Final assembly is by gluing the pieces together to the plan shown in Fig. 4.

Finish the chair by painting it, first with an undercoat then with two coats of polyurethane gloss. The finished product is both attractive and sturdy.

Fold~down table with cabinet

The design

The table and cabinet consist mainly of ¾in (19mm) birch plywood, all cut from one standard-sized 4ft × 8ft (1219mm × 2438mm) sheet as shown in Fig. 1. Parts A and B are shaped to fit together and joined with strap hinges, then screwed to the wall to form the table when A is dropped down, as shown in Fig. 5. Panel A drops down to rest on battens as shown in Fig. 2. The cabinet is screwed firmly to panel B.

A small bullet catch is fitted in the top edge of panel B, and the cup for this is at the point where panel A butts against it. This holds panel A in position when it is folded against the wall out of the way.

All the cabinet joints are simply butted, glued and screwed. If the unit is to be painted, the screw heads are counter-sunk flush with the surface and filled over with a commercial cellulose filler, which is then sanded down to give a perfectly smooth finish. But if you want to stain or varnish the unit, the screw heads must be sunk deeper and the holes filled with matching boat plugs of wood of a suitable size. When these wood plugs are firmly in place, they must be sanded flush for a smooth, finished look.

Fold-down table with cabinet

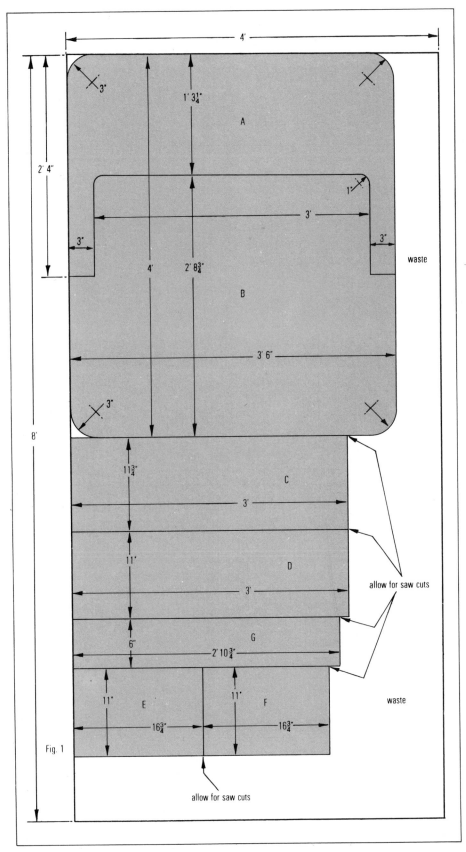

waste

waste

Fig. 1

allow for saw cuts

allow for saw cuts

thick; two large 12in (305mm) strap hinges with mounting screws; one bullet catch no wider than $\frac{5}{8}$in (15mm); two dozen $1\frac{1}{2}$in (38mm) No.8 flathead screws; and some carpenter's glue. You will also need paint or varnish; in the case of the latter you will need some dowelling to make plugs.

For tools, you will need a power circular saw and a jigsaw for separating panels A and B (so you might as well use this for cutting all the pieces); a power drill; jack plane; screwdriver; spoke shave; rule; try square; and an ordinary compass.

Cutting out

Mark out the large sheet of plywood using the dimensions in Fig. 1 as a guide. To ensure that the table top will fit properly over the top of the cabinet when dropped down, it is essential that you begin marking in a particular order.

First mark out panel C. This panel has two square corners, and two with 3in (76mm) radiuses. The method of marking these is shown in Fig. 3. Mark two points 3in (76mm) along each edge, from the corner. Link these two points with lines drawn at right angles to the edges to make a 3in (76mm) square. The inner corner of this square will be the center for the curve, which should be drawn with the compass.

Using the same radius method where rounded corners are required, mark out the rest of the ply sheet, but not the dividing line that will separate panels A and B. Make straight cuts with a circular saw and straightedge.

With the jigsaw, carefully cut round the outline of panel C and when this has been done, finish the cut edges of the panel with the jack plane (fine set) along the straight edges, and the spokeshave at the rounded corners or use a belt sander.

Mark out the dividing line between panels A and B, using panel C as a template for the central piece. This is done to ensure that when the table top is dropped, it will fit exactly over the top of the cabinet, which will be panel C.

All the remaining panels can now be cut out of the large sheet of ply. Finish off all the cut edges of panels A and B, but leave the remainder of the panels, which will form the cabinet, until you assemble them.

Making the cabinet

This is constructed from panels C, D, E,

Materials and tools

The unit will require one 4ft × 8ft (1219 mm × 2438mm) sheet of $\frac{3}{4}$in (19mm) birch plywood (pick a piece with two good sides); one length of $1\frac{1}{2}$in × 3ft (38mm × 914mm) hardwood $\frac{1}{2}$in (13mm)

F and G, fastened together with glued and screwed butt joints.

Trial assemble the cabinet panels as shown in Fig. 2 to check that all the edges butt neatly. Where they do not, plane the high spots down. Then finish off all the edges that will be visible at the front of the unit. Trial assemble again to ensure that the cabinet is square and that all parts fit correctly.

Drill screw holes at 2in (51mm) intervals through the ends of the top and bottom panels C and D. Spread adhesive along the top and bottom of the side panels (E and F), then assemble the cabinet in the shape shown in Fig. 2, and drive the screws home into the side panels.

Before the glue has set, lightly mark a horizontal line with a try square halfway down the outside of each side panel and drill screw holes along the lines at 2in (51mm) intervals. Trim panel G to fit in between the sidepanels, spread adhesive along each of its edges, place it in position and drive the screws home.

Check that the cabinet is square, then leave it while the glue sets.

Assembling the unit

This is relatively simple except that care is required to make the table leaf drop accurately over the cabinet top.

Lay panels A and B down on a flat surface and fit the strap hinges (Fig. 4) to link them as shown in Fig. 5. Lay the cabinet on its back, on top of panel B, in the approximate position where it will eventually be fixed. Lift the table leaf up to a vertical position and adjust the position of the cabinet until the top fits neatly into the recess at the base of the leaf. When the cabinet is in the right position, lightly mark a line around its top, base and sides onto the surface of panel B.

Take the cabinet off, then drill screw holes at 2in (51mm) intervals around the inside of the marked line. Each hole should be drilled inside the line at a distance equal to half the thickness of the plywood used for the cabinet body. Lay sections A and B on a table, place the cabinet in position again, ease part of panel B over the edge of the table to expose some of the screw holes underneath, then drive several screws in. Repeat this until the cabinet has been screwed all around and is securely fixed to panel B.

Finally, fit the leaf stop. This is the 3ft (914mm) length of hardwood. Place

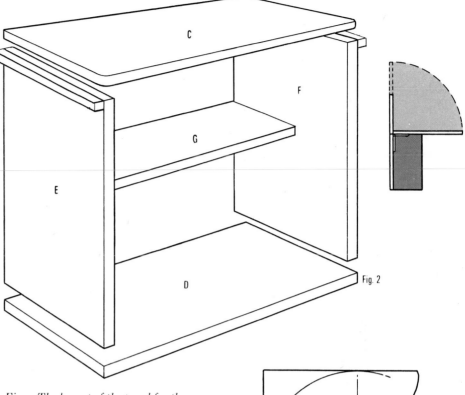

Fig 1. The layout of the panel for the fold-down table.

Fig 2. The construction of the cabinet.

Fig 3. The method used to mark the corner radii.

Fig 4. The back panel and table top are hinged with strap hinges.

Fig 5. The completed back panel assembly. The table is held in the closed position by a bullet catch at the top of the lower panel.

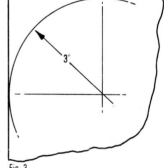

Fig. 3

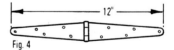

Fig. 4

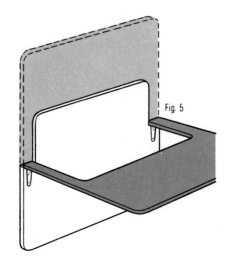

Fig. 5

it in position underneath panel C so that one third of its width protrudes beyond the front of the panel. Mark a line along the front edge of C on the leaf stop. Remove the stop and drill holes at 2in (51mm) intervals at a point half-way between this marked line and the back of the stop. Replace the stop and screw it in position.

Finishing

Fill any cracks or holes with filler, allow it to dry, then rub the surface down.

To avoid getting any paint on the walls, it is best to paint the unit before fixing it to the wall. When the paint is dry, all that remains is to fix the unit to the wall with screws and wall plugs.

Made~to~measure kitchen unit

Commercially available kitchen units are expensive and may not always fit the general layout of your kitchen. This made-to-measure unit can be adapted to your own particular requirements.

The wide, corner sub-assembly, clearly illustrated in Fig. 1, is ideal for a work surface adjacent to the cooking area. The rest of the unit is fitted with drawers or shelving and is designed to contain a sink recessed into the work surface. The advantages of this system are clear. Food can be prepared next to the cooking area; and dirty dishes can be stacked on one side of the sink and, when washed, placed on the other side to drain. All washing materials and utensils can be stored in the adequate drawers and recesses, for immediate access. In conjunction with a wall-mounted unit for holding foodstuffs, you have a complete and labor-saving system for your kitchen.

Obviously, your own kitchen may require a modification of this design, but this can be achieved relatively easily. Comprising two sub-assembles linked by a work surface, the kitchen unit can be built to incorporate say, a dishwasher or other item, and the dimensions can be altered with no modification of construction techniques. Perhaps the major obstacle in a particular kitchen would involve certain alterations in the unit design, but the unit is so adaptable and simple that the resourceful carpenter can overcome this obstacle with only a little extra work.

As stated above, there is no reason why the dimensions of the unit cannot be altered to individual requirements. The length of the unit is limited only by the space available; its depth can be adjusted if, for example, you wish to incorporate a dishwasher into a recess. The height of the unit, as shown on the plan, is 34 in (864mm) which is suitable for the smaller person. Standard kitchen furniture is generally 36in (914mm) high; but the height of the unit can be adjusted to suit your own requirements.

Preliminaries

Before embarking on the construction of a large piece of furniture like the kitchen unit, you are advised to bear several factors in mind. When assembled, this unit will not pass through conventional doorways so, unless you have the freedom of the kitchen for a few days, you must temporarily assemble the unit in the workshop and, when all cutting is complete, transfer it for final assembly to the proposed site. As the unit comprises a large number of individual members, you are advised to consult the cutting list and group similar materials into the same sizes where possible. This makes the job of ordering the materials much easier and prevents waste.

The materials

You can choose the materials for the unit to match the style of your existing kitchen. Those given have been chosen for their good looks, economy and hard-wearing qualities, but you can make suitable substitutions. Thus, while $\frac{3}{4}$in (19mm) plywood is recommended for the main work surface, an alternative would be hardboard. Plywood or plain hardboard should be covered with a hard-wearing material. As well as being used for the work top, plywood is recommended for the partitions. All other panels, including the doors and drawers, are cut from $\frac{5}{8}$in (16mm) thick plywood.

Frame members, which include the ground-level kickboards, are of softwood dressed on four sides. The backs of both sub-assemblies are made from hardboard, as are the drawer bottoms which are covered by a veneer. All the drawer battens and stops are cut from a hardwood such as oak.

Construction

When making a large unit comprising many parts, it is best to work in a strict order of procedure. You can either build each sub-assembly separately, or you can cut out and assemble similar parts of the sub-assemblies together. Remember, though if you are making the wide corner sub-unit, it must be constructed either on the site, or trial assembled elsewhere.

Begin by cutting out the four main partitions for the sink unit to the plans in Figs. 2 and 3, using an electric circular saw with combination blade. Cut the cut-outs in the partitions with a jig-saw. The two ends have cut-outs made at the top corners to house the two upper long rails. Another cut-out, at the front lower corner, houses the kickboard. Cut-outs similar in size and location to those on the top edge of the ends are made on the two intermediate partitions; but the lower edges of these pieces are cut to a different plan which is shown in Fig. 2. If the unit is to house a standard size kitchen sink, you will have to modify the cabinet to accommodate it. The exact size and shape of such a cut-out will depend on the particular design of your own sink.

Now cut out the lower shelf and then the two top 1in x 3in (25mm x 76mm) softwood rails to the sizes given in the cutting list. Lay the rails on top of the shelf so that they are parallel to its long edges and are $\frac{3}{4}$in (19mm) from each end, and mark the position of each partition as given in Fig. 1.

Screw and glue 1in x 1in (25mm x 25mm) jointing members to the inside surface of the two ends. Their exact location is shown in Fig. 1.

At this stage, before beginning assembly, add veneer or laminate strips to all the exposed front edges of the partitions and lower shelf.

With all exposed edges laminated, fix the intermediate partitions to the lower shelf by gluing and screwing, (through the shelf). Then turn the unit over and add the top rails, referring to the previously marked positions of the partitions to give their exact location.

Before fixing the ends, add the hardboard back panel by nailing and gluing it to the rear edges of the intermediate partitions and the top, rear rail. Now that the addition of the back panel has given greater stability to the unit, you can glue and screw the ends into position. Check at this stage that the pieces are square to each other.

Turn the unit right side up and fix the kickboard into position. This piece is recessed into the bases of the ends and extends the whole length of the sub-assembly. It is held in position by being fixed to a 1in x 1$\frac{3}{8}$in (25mm x 35mm) batten which is glued and screwed to the underside of the lower shelf. When assembled correctly this member is over lapped by the doors. As added support wood reinforcement brackets (cut to the shape shown in Fig. 4) are screwed into the internal right angles made by the kickboard/bottom shelf joint. There are four of these brackets and each is located at a corner of the unit.

Shelving

You can choose shelving to suit your individual needs. In this unit the sink is recessed onto only one of the partitions, and the cabinet space that it occupies, i.e. the cabinet formed by the end and first partitions, is left clear so that it can store large items such as buckets and washbasins. The space between the partitions is fitted with a single shelf situated about halfway up the cabinet.

Made-to-measure kitchen unit

Fig 1. Exploded view of the fitted kitchen unit.

Fig 2. The end panels of the unit. These are the only two panels which are full depth.

Fig 3. The intermediate panels finish at the unit base. The dotted line shows the cutout which must be made to accommodate the sink unit.

Fig 4. The brackets supporting the kickboard are clearly illustrated, as are the top rails and the position of the shelf.

Fig 5. Front view of the end unit showing position of the drawers and runners.

Fig 6. Plan view of the end unit.

Fig 7. Side elevation of the end unit.

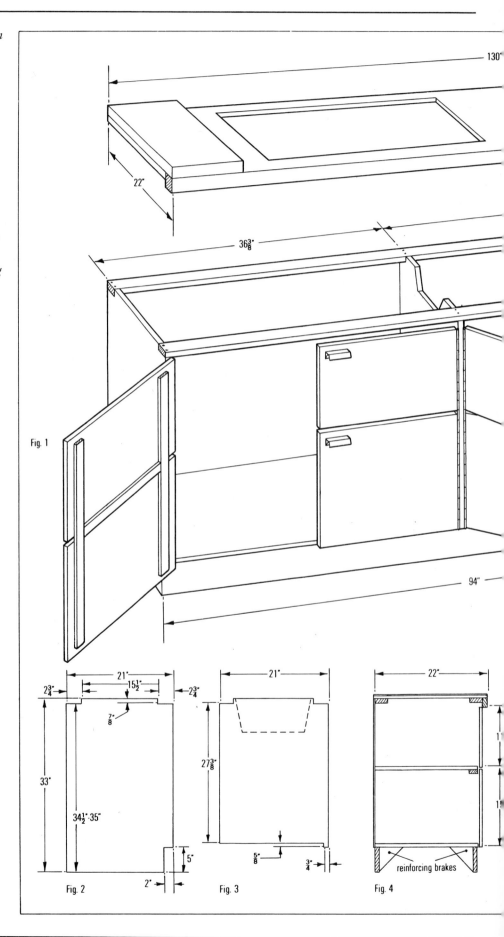

Fig. 1

Fig. 2

Fig. 3

Fig. 4

reinforcing brakes

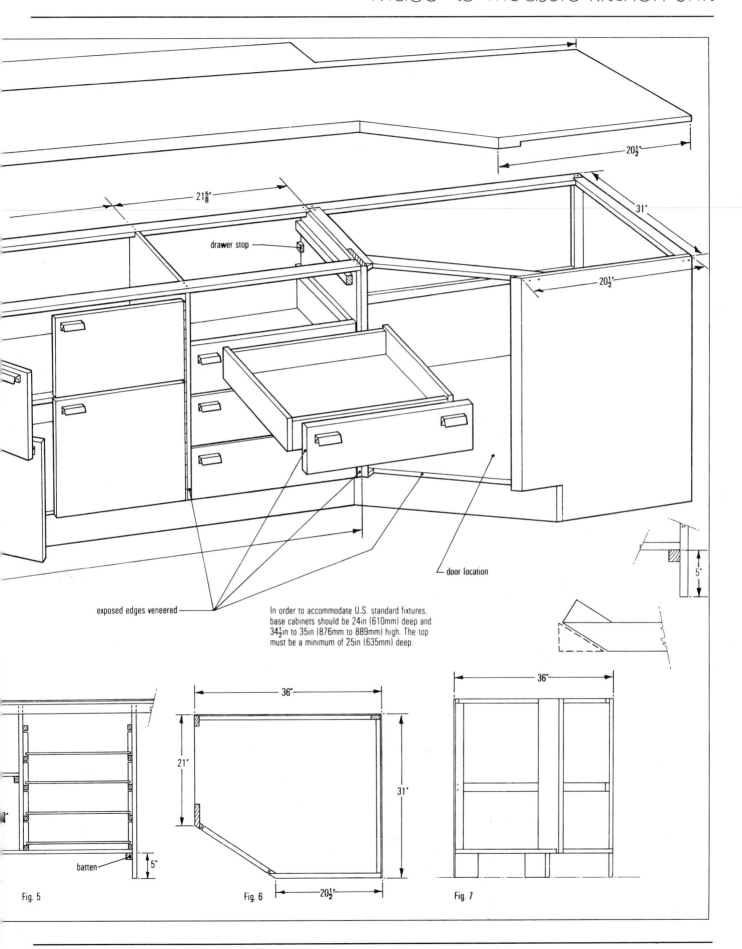

drawer stop

$21\frac{5}{8}"$

$20\frac{1}{2}"$

31"

$20\frac{1}{2}"$

$20\frac{1}{2}"$

5"

door location

exposed edges veneered

In order to accommodate U.S. standard fixtures, base cabinets should be 24in (610mm) deep and $34\frac{1}{2}$in to 35in (876mm to 889mm) high. The top must be a minimum of 25in (635mm) deep.

batten

5"

Fig. 5

36"

21"

31"

$20\frac{1}{2}"$

Fig. 6

36"

Fig. 7

Made~to~measure kitchen unit

The space between the second partition and the other end is used for four drawers. Obviously you can modify this arrangement as necessary.

The easiest method of installing the shelves is to simply rest them on 1in x 1in (25mm x 25mm) battens glued and screwed to the partitions. This method allows you to remove shelves for cleaning. Under the weight of heavy kitchen utensils, even a hardboard or plywood shelf will bend. To prevent this, glue and screw a 1in x 2in (25mm x 51mm) batten under each shelf, flush with the front edge as shown in Fig. 4.

Drawer construction

In a unit of this size, there is a possibility that the finished dimensions will deviate slightly from those given in the cutting list and Fig. 1. As a result, you are advised to use the dimensions of the drawer members given in the cutting lists as a guide, and to take the exact measurements for the drawers from the unit itself. Aim for a $\frac{1}{8}$in (3mm) clearance between the drawer sides and the partitions.

Make up the drawers to the plan shown in Fig. 1, after cutting out the pieces for a good fit. The two center drawers are larger than the top and bottom drawers. All the drawer bottoms are cut from laminate-covered hardboard and fit into $\frac{1}{4}$in (6mm) deep grooves cut into the sides $\frac{1}{4}$in (6mm) above their bottom edges. The sides are simply butt-jointed to the front and back and secured with glue and $1\frac{1}{2}$in (38mm) finishing nails. A false or decorative front, cut from plywood is added to each drawer front panel, but before fixing this piece in position, check that the main drawer bodies fit satisfactorily.

To do this, glue and screw the lower pair of hardwood drawer runners into position on the lower shelf, flush with the inside surfaces of the partitions. Slide one drawer into position, slip a piece of cardboard onto its upper edges to give the correct clearance, then mark out the position of the next pair of drawer runners and fix them into position. Repeat this procedure with the other two drawers and, when you are satisfied that each drawer fits well, add the false drawer fronts.

Note that the lower drawer front is wide enough to extend below the bottom of the drawer to cover the main unit lower shelf edge. When fitted correctly, the other drawer fronts should just have a working clearance between each other, and the top drawer front must clear the front edge of the working top.

Door construction

The dimensions of each door will depend on the arrangement of the internal shelving. Door sizes given in the cutting list are ideal for the unit shown. Having cut out all the door panels to size, check for fit, remembering that each panel is hung on the inside faces of the partitions on full-length, piano-type hinges. Where internal shelving is installed, (as in the center cabinet, shown in Fig. 4), upper and lower door panels are installed separately. where no shelf is installed, door panels should be joined by $\frac{3}{4}$in or $\frac{1}{2}$in (19mm or 13mm) battens, glued and screwed to the rear surfaces, as shown in Fig. 1.

Cut the piano-type hinges to length, and then screw them into position on the edges of the doors. Mark the correct position of the hinges on the partitions and screw the doors to the unit. Finally, fit each door with a magnetic-type catch and add the handles of your choice.

Corner sub-assembly construction

As stated previously, the dimensions and shape of this sub-unit will depend on individual preference and the existing scheme of your kitchen. Remember, a unit of this size is too large to fit pre-assembled through the average kitchen door.

If you intend to make the sub-unit as shown, cut the lower main shelf to the shape and size shown in Fig. 6. All the corner unit parts are assembled using simple butt or lap joints, but you may experience some difficulty in cutting out the two panels which have angled long edges. Having examined Fig. 1 to discover the precise arrangement of the various members, use direct measurement to mark up the pieces. To angle the edges of the panels, first mark out the correct angle on the upper and lower edges of each. Then, using a marking gauge and pencil, draw a line along the inside surface of each panel between the marked points. Cut the wood down to this line. Do not discard the off-cut; it is used to provide a right-angled edge on which the door is hung, as shown in the detail in Fig. 1. When you have cut out

the panels, fix them to the base.

Build up the rest of the frame from 1in x 2in (25mm x 51mm) softwood. The two rear vertical members simply butt on to the underside of the top rails which are joined in a modified lap joint.

Add the kick board and the support blocks and then the two hardboard back panels. Cut the door to size and hang it between the two angled off-cuts as described previously.

Working top construction

Measure the working top against the assembled units. If you wish to incorporate a recess between the sub-assemblies you must allow for this. The dimensions of the top given in the cutting list apply only if the two sub-units are butted together.

Glue and screw the 1in x 1¼in (25mm x 32mm) softwood, front edge thickener under the top, so that its front edge is flush with the front edge of the working top. Then cover the whole surface with a plastic laminate, using a suitable adhesive. Add a hardwood work block to the surface in a convenient position.

Now you must cut out a space in the work top to receive the sink and its surround. Assuming that you want to fit a new sink with an integral drainboard, you must make sure that its dimensions will fit the top of the unit. Most sink/draining units come complete with fixing instructions and, provided you make a watertight seal between sink and worktop, you should experience little difficulty in fitting a sink unit. You can cut the necessary hole using a jigsaw. The area of wall behind the sink will have to be covered with a splash-proof surface of some kind. One of the most attractive materials for this purpose is ceramic tiling. The only other modification to the work top that may be required is the boring of holes large enough to take taps.

When you have checked the precise location of the sink in relation to the unit and have cut away the necessary areas of the worktop, it can be fixed into position by gluing and screwing it to the horizontal main battens—driving the screws through the battens into the top. Do any plumbing and sealing now.

Although the construction of the unit will have entailed a lot of work, the finished result more than justifies it. The kitchen unit will greatly save time in the kitchen, aid hygiene and introduce a contemporary touch.

Cutting list: Corner sub-unit

Wood	standard	metric
1 veneered lower shelf	$\frac{5}{8}$ x 31 x 36	16 x 787 x 914
1 angled panel	$\frac{3}{4}$ x 22 x 33	19 x 559 x 838
1 angled panel	$\frac{3}{4}$ x 5 x 28	19 x 127 x 711
2 softwood vertical frame members	1 x 2 x 26$\frac{3}{8}$	25 x 51 x 670
1 softwood horizontal frame member	1 x 2 x 36	25 x 51 x 914
1 softwood horizontal frame member	1 x 2 x 34	25 x 51 x 863
2 softwood horizontal frame members	1 x 2 x 29$\frac{1}{4}$	25 x 51 x 743
1 softwood horizontal frame member	1 x 2 x 21	25 x 51 x 533
1 softwood horizontal frame member	1 x 2 x 22	25 x 51 x 559
1 softwood jointing batten	1 x 1 x 17	25 x 25 x 431
1 veneered kickboard	$\frac{5}{8}$ x 5 x 22	16 x 127 x 559
1 veneered lower door	$\frac{5}{8}$ x 15 x 16$\frac{1}{4}$	16 x 381 x 413
1 veneered upper door	$\frac{5}{8}$ x 11$\frac{1}{2}$ x 16$\frac{1}{4}$	16 x 292 x 413
4 support blocks	$\frac{3}{4}$ x 5 x 5	19 x 127 x 127
2 hardwood door jointing battens	$\frac{1}{2}$ x $\frac{3}{4}$ x 26	13 x 19 x 660
1 hardwood door jointing batten	$\frac{1}{2}$ x $\frac{3}{4}$ x 13	13 x 19 x 330

Cutting list: Work top

Wood	standard	metric
1 plywood or hardboard top	$\frac{3}{4}$ x 31 x 130	19 x 788 x 3300
1 hardboard working top block	$\frac{3}{4}$ x 12 x 22	19 x 305 x 559
1 softwood edging strip	1 x 1$\frac{1}{4}$ x 114	25 x 31 x 2896
2 softwood edging strips	1 x 1$\frac{1}{4}$ x 22	25 x 31 x 559
2 softwood edging strips	1 x 1$\frac{1}{4}$ x 10	25 x 31 x 254

You will also require:

Plastic laminate, $\frac{5}{8}$in (16mm) veneer strips. Door handles. Wood-working glue. 144 1$\frac{1}{2}$in (38mm) finishing nails and wood screws. Laminate adhesive.

Cutting list: Main sink unit

Wood	standard	metric
2 ends	$\frac{3}{4}$ x 21 x 33	19 x 533 x 838
2 intermediate partitions	$\frac{3}{4}$ x 21 x 28	19 x 533 x 711
1 veneered lower shelf	$\frac{5}{8}$ x 20$\frac{1}{4}$ x 92$\frac{1}{2}$	16 x 514 x 2349
1 veneered center shelf	$\frac{5}{8}$ x 20$\frac{1}{4}$ x 35$\frac{1}{4}$	16 x 514 x 895
4 veneered upper doors	$\frac{5}{8}$ x 11$\frac{1}{2}$ x 17$\frac{1}{2}$	16 x 292 x 445
4 veneered lower doors	$\frac{5}{8}$ x 15 x 17$\frac{1}{2}$	16 x 381 x 445
8 veneered drawer sides	$\frac{5}{8}$ x 5$\frac{1}{2}$ x 20	16 x 140 x 508
8 veneered drawer ends	$\frac{5}{8}$ x 4$\frac{1}{2}$ x 19$\frac{1}{4}$	16 x 114 x 489
2 veneered drawer fronts	$\frac{5}{8}$ x 7$\frac{1}{4}$ x 20$\frac{1}{2}$	16 x 184 x 521
2 veneered drawer fronts	$\frac{5}{8}$ x 6$\frac{1}{8}$ x 20$\frac{1}{2}$	16 x 156 x 521
1 veneered kickboard	$\frac{5}{8}$ x 5 x 95	16 x 127 x 2413
2 rear supports	$\frac{3}{4}$ x 5 x 5	19 x 127 x 127
4 kickboard reinforcements	$\frac{3}{4}$ x 5 x 5	19 x 127 x 127
2 softwood top main battens	1 x 3 x 90	25 x 76 x 2286
1 softwood shelf front edge	1 x 2 x 35$\frac{1}{4}$	25 x 51 x 895
10 hardwood drawer runners	$\frac{1}{2}$ x $\frac{3}{4}$ x 20	13 x 19 x 508
4 hardwood door jointing battens	$\frac{1}{2}$ x $\frac{3}{4}$ x 26	13 x 19 x 660
2 hardwood door jointing battens	$\frac{1}{2}$ x $\frac{3}{4}$ x 13	13 x 19 x 330
4 laminated hardboard drawer bases	19$\frac{3}{4}$ x 20	502 x 508
4 hardwood jointing and shelf supports	1 x 1 x 19	25 x 25 x 483
1 hardboard back	$\frac{1}{8}$ x 28 x 94	3 x 711 x 2388

Split level oven/cook top

Above: The attractive lines of the split-level oven and cooktop unit are shown here to good advantage. Some adjustments to the dimensions given here may be required to accommodate the units used and the size of your particular kitchen.

This kitchen unit comprises a recessed sink and cooktop unit which butts onto an eye-level oven unit.

Here, again, as in the previous projects, the basic material to use for the body and both sub assemblies, is flakeboard covered by plastic laminate. There are other materials to use, but whatever type you choose, ensure that the top covering is heat resistant. Plywood or flakeboard may be used for the working top—again, when choosing a covering, check that it is heat resistant.

You may wish to modify the dimen-

sions of the unit to suit your kitchen; the simplicity of the design allows this. One cautionary word, however, before you start construction—check the oven and cooktop manufacturer's instructions and specifications and change the design shown where necessary.

The cooktop/sink unit
Use an electric jigsaw with a fine 3in (76mm) blade for the cutting. A straight-edge will help you cut accurate, straight pieces. Cut all the long sections with a circular saw. Begin by cutting out the

two ends to the outline shown in Fig. 2. These have cut-outs at the top corners to accommodate the two upper long rails. Another cut-out, at the front lower corner houses the kick board. Now cut out the three intermediate partitions to the shape represented by the dotted line in Fig. 2. Again, these pieces have cut-outs similar to those made on the top edges of the end formers but, as they rest flush on the bottom shelf, they do not house the kickboard. Having decided on and checked the locations of the fitted cook-top and sink, make cut outs in the intermediate formers to accommodate these fixtures.

Next cut out the lower shelf and then the two 1in x 2in (25mm x 51mm) upper long rails to the sizes given in the cutting list. Lay the rails on top of the shelf so that they are parallel to its long edges and overlap both short edges by $\frac{5}{8}$in (16mm). Mark in the position of each former on the rails, and shelf, as shown in Fig. 1.

Turn the intermediate partitions upside down and, referring to the previously marked locations, nail the lower shelf to their bottom short edges. Secure the partitions with 1$\frac{1}{2}$in (38mm) No. 6 screws. Set the structure right-side up and add the the long rails into the appropriate housings. At this point, add the end partitions by nailing and screwing and, when you have checked that the unit is square, fit the hardboard back panel.

Now you can add the kick board. First fix a full-length 1in x 1in (25mm x 25mm) softwood batten to the underside of the lower shelf at the location shown in Fig. 2. The kickboard butts onto this member and fits between the end partitions. Screw it with 1$\frac{1}{2}$in (38mm) No.6 screws.

The top front panel is butt-jointed to the front edges of the partitions and ends at the location shown in Figs. 1 and 2.

Shelving

You can choose shelving to suit your individual needs, remembering that the recessed sink and cooktop will not leave very much space under them. The simplest method of installing any shelves is to rest them on 1in x 1in (25mm x 25mm) hardwood battens glued and screwed to the partitions. This method allows you to remove shelves for cleaning. Remember that under the weight of heavy kitchen utensils even flakeboard can bend. To prevent this, glue and screw a 1in x 2in (25mm x 51mm) batten under each shelf,

flush with the front edge to add rigidity.

Once any shelving is in place, you must effectively fireproof the area around the cooktop location. Different types require different treatment, so before starting work consult the manufacturer's instructions or, if you do not possess these, the manufacturers themselves. Asbestos is the most effective fire-proofing material in common use. Cut to size with a fine-toothed saw and fix it to those areas where it is needed.

Drawer construction

Drawers are used at one end of the unit only. When marking out the individual drawer panels, use the dimensions given in the cutting list as a guide and take the exact measurements from the unit itself. Aim for a $\frac{1}{8}$in (3mm) clearance between the drawer sides and the partitions.

When you have cut out each piece, make $\frac{1}{4}$in (6mm) deep grooves in the drawer sides to accommodate the drawer bottoms. The bottom edges of the grooves are located $\frac{1}{4}$in (6mm) from the bottom edges of the side panels. Once the drawer bottom has been glued into position, add the front and back panels which simply butt between the sides. Secure them by driving screws through the bottom panel into their lower edges.

An additional false front is added to each drawer, but before putting this piece in position, check that the main drawer bodies fit satisfactorily. To do this, glue and screw a pair of hardwood drawer runners into position on the lower shelf, flush with the inside surfaces of the partitions. Slide one drawer into position, slip a piece of cardboard onto its upper edges to give the correct clearance, then mark out the position of the next pair of drawer runners and fit them into position. Repeat this procedure with the other two drawers and, when you are satisfied that each drawer fits well, add the false fronts.

Door construction

Having cut out all the door panels to size, check for a good fit, remembering that each panel is hung on the inside faces of the formers on full-length, piano-type hinges. Cut these hinges to length and then screw them into position on the edges of the doors. Mark the correct position of the hinges on the partitions and end panels and screw the doors to the unit. Finally, fit each door with a magnetic-type catch and add handles.

Split level oven/cooktop

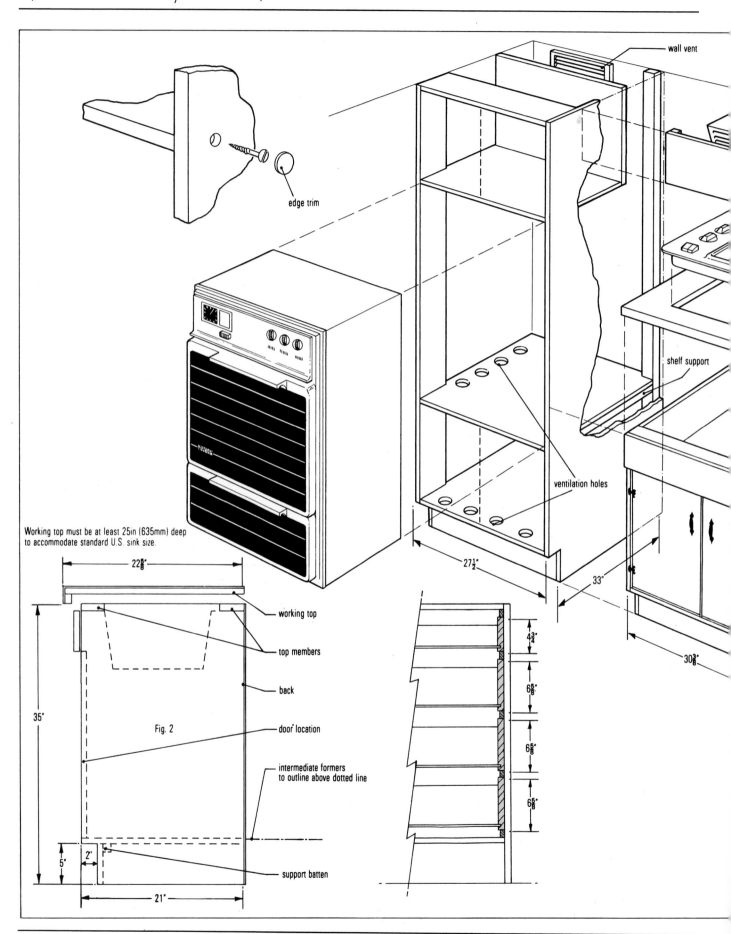

wall vent

edge trim

shelf support

ventilation holes

Working top must be at least 25in (635mm) deep to accommodate standard U.S. sink size.

working top

top members

back

door location

intermediate formers to outline above dotted line

support batten

Fig. 2

22⅝"

35"

5"

2"

21"

27½"

33"

30⅜"

4¾"

6⅝"

6⅝"

6⅝"

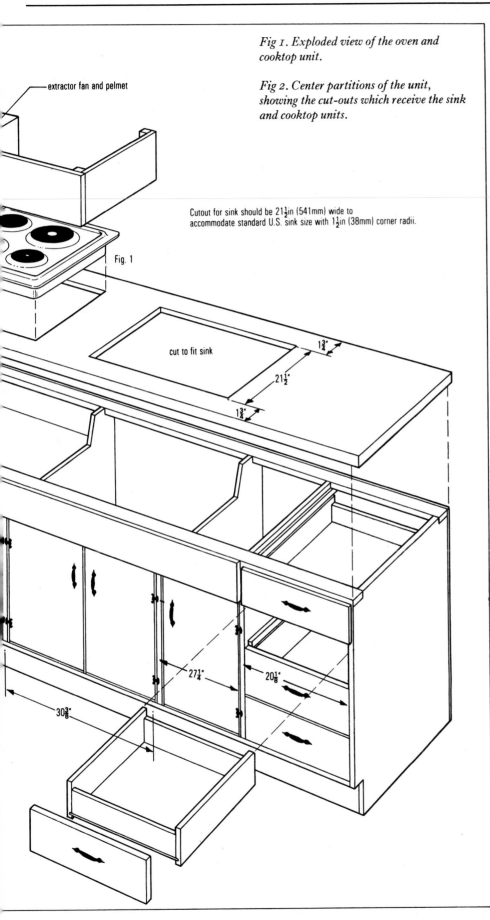

Fig 1. Exploded view of the oven and cooktop unit.

Fig 2. Center partitions of the unit, showing the cut-outs which receive the sink and cooktop units.

extractor fan and pelmet

Cutout for sink should be 21½in (541mm) wide to accommodate standard U.S. sink size with 1½in (38mm) corner radii.

Fig. 1

cut to fit sink

1¾"

21½"

1¾"

27¼"

20⅛"

30⅜"

Working top construction

Cut out the working top and glue and screw a 1in x 1in (25mm x 25mm) soft-wood strip to the front and right hand edge of this piece, as shown in Fig. 2. Then cover the surface with a plastic laminate. If desired, add a hardwood work block to the surface in a convenient position.

Now cut out spaces for the sink and cooktop to the manufacturers' specifications. When you have made the cut-outs, fix the working top in position by gluing and screwing it to the upper long rails—driving the screws through the rails into the top. Have any plumbing or electrical work carried out at this point and then cover the area of wall immediately behind the sink with a splash-proof material, such as ceramic tiles.

Making the oven unit

Before starting to build the unit, you must check the dimensions of your oven against the dimensions of the unit shown and, if necessary, modify both the design and sizes of the various parts of the unit. Then cut out the two side members and the shelves. Take the two lower shelves and make cut-outs at one corner to accommodate the narrower side member. These shelves also house the rear 1in x 2in (25mm x 51mm) support batten and cut-outs must be made in their rear edges to accommodate this member. The top shelf also fits around the narrow side member, but butts directly onto the support batten. Before securing this shelf to the main structure, glue and screw the back panel to it.

With all the pieces cut out, begin assembly by gluing and screwing the side members to the shelves. Add the recessed kickboard and the top cross member. Next, stand the unit on its proposed site and mark out on the wall the locations of the support battens. Fix these in place, using wall plugs, and screw the unit to them. Before fitting the oven in place on the center shelf, add 1in x 1in (25mm x 25mm) reinforcing battens as shown in Fig. 1. Then bore out ventilation holes in both the bottom and center shelves.

Doors are fitted in the same way as described for the sink/cooktop unit. They should be equipped with handles to match the sink unit. If you intend fitting extractor fans above the oven and cooktop, you can increase their efficiency by building a simple cover that adjoins the oven unit. The method of construction for this cover is clearly illustrated in Fig. 1.

In order to achieve a good finish, all exposed screw heads can be covered by small laminate discs cut from matching iron-on edge trim. Any exposed door edges must also be trimmed.

The construction and fitting of this unit will transform your kitchen and make it a more pleasant place to work in.

Once you have experienced the advantages of this system, you may wish to make matching units to hold foodstuffs and kitchen equipment. By using the basic designs described in this chapter, this should present no problem, and the resulting scheme in your kitchen will more than match expensive, manufactured kitchen systems both in quality and appearance.

Sink/Cooktop unit: Cutting list

Wood	standard	metric
Laminated flakeboard or plywood		
2 ends	$\frac{5}{8} \times 21 \times 35$	16 × 533 × 889
3 intermediate partitions	$\frac{5}{8} \times 21 \times 30\frac{3}{8}$	16 × 533 × 772
1 lower shelf	$\frac{5}{8} \times 21 \times 96$	16 × 533 × 2438
5 doors	$\frac{5}{8} \times 15 \times 25\frac{1}{8}$	16 × 381 × 638
1 top drawer front	$\frac{5}{8} \times 6\frac{1}{4} \times 18$	16 × 159 × 457
2 center drawer fronts	$\frac{5}{8} \times 7\frac{1}{4} \times 18$	16 × 184 × 457
1 bottom drawer front	$\frac{5}{8} \times 7\frac{3}{4} \times 18$	16 × 197 × 457
1 top front panel	$\frac{5}{8} \times 7 \times 77\frac{1}{4}$	16 × 178 × 1962
1 kickboard	$\frac{5}{8} \times 5 \times 96$	16 × 127 × 2438
Hardwood or flakeboard		
1 working top	$\frac{3}{4} \times 22\frac{5}{8} \times 98\frac{1}{4}$	19 × 575 × 2496
Softwood		
2 top rails	$1 \times 2 \times 97\frac{1}{4}$	25 × 51 × 2470
1 front edge	$1 \times 1 \times 98\frac{1}{4}$	25 × 25 × 2496
1 front edge	$1 \times 1 \times 21\frac{5}{8}$	25 × 25 × 549
Hardboard		
1 back	$\frac{1}{8} \times 35 \times 97\frac{1}{4}$	3 × 889 × 2467
4 drawer bottoms	$\frac{1}{8} \times 17\frac{1}{2} \times 19$	3 × 445 × 482
Plywood		
6 drawer sides	$\frac{1}{2} \times 6\frac{5}{8} \times 19$	13 × 168 × 482
2 top drawer sides	$\frac{1}{2} \times 4\frac{3}{4} \times 19$	13 × 121 × 482
2 top drawer ends	$\frac{1}{2} \times 3\frac{1}{2} \times 17$	13 × 89 × 432
6 drawer ends	$\frac{1}{2} \times 5 \times 17$	13 × 127 × 432
Hardwood		
10 drawer runners	$\frac{1}{2} \times \frac{3}{4} \times 20$	13 × 19 × 508

You will also require:

Plastic laminate. Fireproofing material. Door handles. Woodworking glue. 1 gross $1\frac{1}{2}$in (38mm) finishing nails and No.6 screws. Edge trim.

Oven unit: Cutting list

Wood	standard	metric
Laminated flakeboard or plywood		
1 side panel	$\frac{5}{8} \times 21 \times 80$	16 × 533 × 2032
1 side panel	$\frac{5}{8} \times 6 \times 80$	16 × 152 × 2032
2 bottom shelves	$\frac{5}{8} \times 21 \times 31$	16 × 533 × 787
1 top cross member	$\frac{5}{8} \times 6 \times 30\frac{3}{4}$	16 × 152 × 781
1 top shelf	$\frac{5}{8} \times 17 \times 31$	16 × 432 × 787
1 top back panel	$\frac{5}{8} \times 24 \times 31$	16 × 610 × 787
2 top doors	$\frac{5}{8} \times 15 \times 16\frac{1}{4}$	16 × 381 × 413
2 bottom doors	$\frac{5}{8} \times 15 \times 18$	16 × 381 × 457
1 kickboard	$\frac{5}{8} \times 5 \times 29\frac{3}{4}$	16 × 127 × 756
Softwood		
1 vertical support batten	$1 \times 2 \times 80$	25 × 51 × 2032
2 center shelf battens	$1 \times 1 \times 20$	25 × 25 × 508

Kitchen wall cabinet

Above: The completed kitchen wall unit in position—basic and adaptable, it can be adjusted to fit into almost any kitchen.

Although kitchens require more storage space than most other rooms, they are often too small to allow adequate standing units to be installed. One satisfactory solution is to build a spacious wall-hung cabinet, purpose-built for your kitchen.

This kitchen wall cabinet or cupboard has been designed to take advantage of the durability of plastic laminate – and the beauty. You can use this material to cover flakeboard and come up with a highly durable unit that cleans with the wipe of a damp cloth. While plastic laminate is not cheap, the flakeboard is; yet the resulting structure will be strong. If desired, the unit can be built of conventional solid wood, but this involves more work in finishing and is not as easy to clean – though you can, through a little more time in the final stages, finish it to be relatively easy to clean.

Kitchen wall cabinet

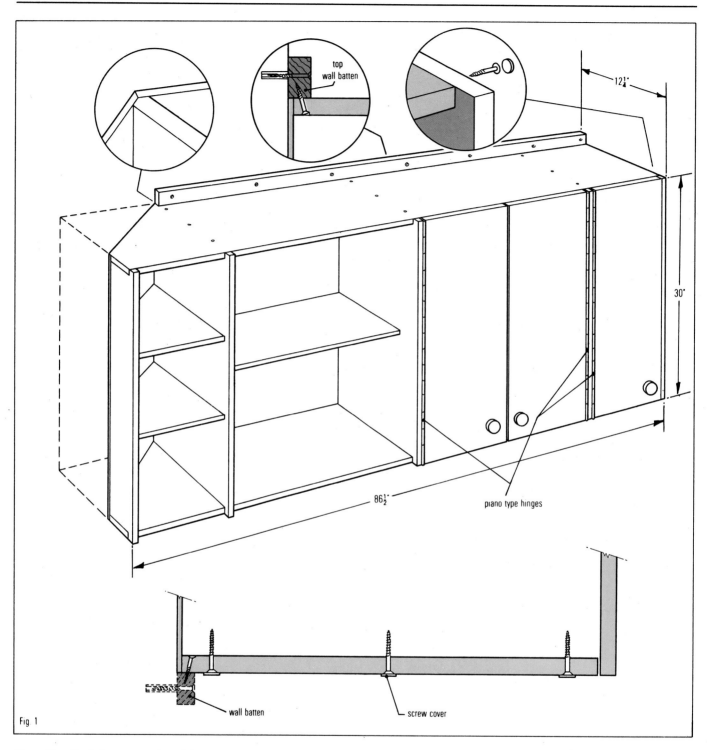

top wall batten

piano type hinges

$86\frac{1}{2}"$

$12\frac{1}{4}"$

30"

wall batten

screw cover

Fig. 1

Fig 1. Details of the construction of the compactly designed kitchen wall-hung cabinet. Both the number and the position of the storage shelves are purely a matter of personal choice and necessity.

The design of the cabinet

The dimensions of the unit are suitable for most kitchens, but can be modified to suit individual requirements. Details of the shelving are not included, but the methods of fixing shelves to the unit are described.

A feature of this cabinet is the angled end, which has been designed to fit a corner of an existing kitchen. Obviously, this angled corner is unsuitable for most kitchens, but details of its construction are included to show the techniques involved in making a fitted cabinet. It is a simple job to modify the design to give a conventional square-ended unit.

Working laminate flakeboard

It is best to cut out all the flakeboard parts needed and then cover them individually with plastic laminate. Pieces of laminate should be cut slightly oversize, then have contact cement applied to both laminate and flakeboard. When the adhesive gets tacky, press the laminate in place, then trim excess with a router fitted with a carbide bit and ball-bearing guide. This will ensure a really clean cut.

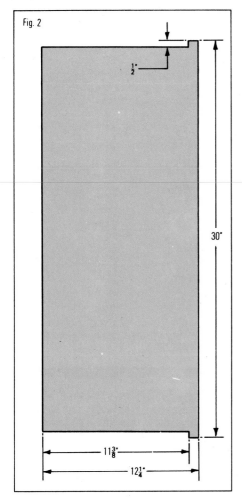

Fig. 2

Fig 2. The dimensions of the panels for the kitchen unit. The ends of the panels are cut away so that the top and bottom fit flush with the full-depth doors.

(289mm) long as shown in Fig. 2.

Cut the recesses carefully and trial assemble the panels to check that all the rear edges of the unit are flush. If you are making the unit with an angled end, cut the rear (long edge) of one vertical panel at the correct angle as shown in Fig. 1.

Now clean up all the internal cut edges by lightly sanding them, and mark the locations of the vertical panels on the top and bottom of the unit. The easiest way to do this is to lay the top onto the bottom, so that all the edges are flush, and then mark both at once.

First mark out the location of the vertical end panels, then work inward. The cabinet is fitted with two single 12in (305mm) doors, located at each end and two double 30in (762mm) doors located in the center. The vertical panels are, therefore, positioned to allow for the width of these doors plus their $\frac{1}{8}$in (3mm) hinge or hinges, and a working clearance of $\frac{1}{16}$in (2mm) between closed doors.

Assembling the unit

Before beginning assembly, lay a sheet of hardboard on the ground to prevent damage to the pieces. Initially, the unit is fitted together using finishing nails and glue, before final fixing with screws. In order to make strong glued joints, first 'key' the laminate with coarse sandpaper where the joints occur, then apply a laminate adhesive.

Use only two nails for each joint, and locate these adjacent to where the screw holes are to be drilled. When the unit has been nailed together, drill for $1\frac{1}{4}$in (32mm) No.8 flathead screws, using a combination drill/countersink bit. It is essential, when drilling laminated materials of this type, that a pilot hole be drilled to take the screw threads. Fix the screws in position and remove any glue left on the exposed surface.

Fitting the back

Roughen the finished surface of the hardboard back where the glued joints occur, and glue and nail it into position. If the cupboard is to incorporate an angled end, the hardboard must be cut to shape and two edges mitered as shown in Fig. 1.

When the back is built, install a $\frac{3}{4}$in $\times$ 2in (19mm $\times$ 51mm) hanging strip above the top, screwing through from inside the unit at intervals of 8in (203mm).

With a wall-hung unit of this type, the hardboard edges do not normally show once the unit is fixed. The lower edge is

Cutting out

The uncomplicated design of the cabinet means that the number of different sections to be cut is kept to the minimum. You should devote care to marking and cutting each piece accurately. Once the cutting is complete, assembly is straightforward, there being no complicated joints or time-consuming finishing processes.

Following the general rules already laid out, mark out and cut all the pieces to the sizes given in the cutting list. The body of the unit is constructed from two 96in (2438mm) lengths of 24in (610mm) wide board cut lengthwise to make two pieces $11\frac{3}{8}$in (289mm) wide and $86\frac{1}{2}$in (2197mm) long for the top and bottom. The remaining two $12\frac{1}{4}$in (311mm) wide boards are cut into five 30in (762mm) long panels to make the vertical partitions and ends. Unless you wish to incorporate the angled or mitered end, cut all these pieces as square-ended panels.

The top and bottom of the unit fit into recesses cut in the vertical panels. These recesses are $\frac{1}{2}$in (13mm) deep and $11\frac{3}{8}$in hidden by a supporting batten and, generally, the top edge is above eye level. However, should it be necessary to disguise the top edge, a length of $\frac{1}{2}$in (13mm) quarter round molding can be fixed after the unit has been mounted.

Only the doors and shelving need to be added; but before fixing these, carefully trim the hardboard back flush with the laminate surfaces and cover the screw heads. These screw covers are cut from edge trim material and are fixed by laying them in place and running a warm iron over them.

The cabinet doors
At least one cut edge of the door panels will expose the flakeboard core and should be covered with an edge trim. Various types of trimming are available, but perhaps the longest lasting is iron-on edge trim. Cut this trim to size, and glue it to all cut edges. Or use wood strips for edges.

It is easier to add the door handles at this stage. These may be chosen according to taste but, when fixing handles which require a hole to be drilled through the door, remember to clamp a waste piece of lumber to the drill exit surface in order to avoid splitting the laminate.

The doors are hung on full-length piano type hinges, which are widely available in plastic or metal. One advantage of this type of hinge over conventional door hinges is that it does not need to be recessed into the edge of the door frame. To fix the hinges in position, first mark out their location on the edge of the door and the vertical panels. Screw the hinge to the door edge after having first predrilled for $\frac{3}{4}$in (19mm) No.4 brass screws. Now hang the doors to the unit using $\frac{1}{2}$in

(13mm) screws.

To complete the construction of the doors, fix 5lb (2.2kg) magnetic cabinet catches to the edge of the top panel. In order to reduce noise when closing the doors, a small rubber disc can be glued adjacent to the catches.

Shelving
Shelving should be fitted according to individual requirements. The ideal width of shelves for this unit is 9in (225mm). Each shelf should be supported on adjustable shelf hardware. The shelves can then be lifted out for cleaning.

Wall fixing
Kitchen cabinets, when filled with utensils or food, are heavy and require adequate support. The best way to provide this support is to fix a full-length batten to the wall on which the unit rests. First decide where the cabinet will be, then fix the batten firmly into position. Rest the cabinet on the batten, but before fixing it to the wall, check that the wall surface is flat and true. If it is not, it will be necessary to shim between the wall and the unit, to prevent distorting the structure as the wall screws are driven home.

When you are satisfied that the cabinet is positioned true to the wall, secure it with screws along the hanging strip. These screws are fixed at 18in (457mm) intervals.

The kitchen wall cabinet is now ready for use. As the need arises, you can build other kitchen units in the same materials, using the same techniques. In this way you can provide a complete and attractive storage system, tailor-made for your own kitchen.

Cutting list

Laminated flake board	standard	metric
1 top panel	$\frac{1}{2}\times11\frac{3}{8}\times86\frac{1}{2}$	$13\times289\times2197$
1 bottom panel	$\frac{1}{2}\times11\frac{3}{8}\times86\frac{1}{2}$	$13\times289\times2197$
5 vertical panels	$\frac{1}{2}\times12\frac{1}{4}\times30$	$13\times311\times762$
4 door panels	$\frac{1}{2}\times15\times30$	$13\times381\times762$
2 door panels	$\frac{1}{2}\times12\times30$	$13\times305\times762$
Hardboard		
1 back panel	$\frac{1}{4}\times30\times86\frac{1}{2}$	$6\times762\times2197$

You will also require:
6 30in (762mm) plastic/metal piano type hinges. 6 door handles. Shelving and shelving battens to individual requirements. Edge trim for door edges. 4 5lb (2.2kg) magnetic catches. 2 full-length supporting battens. Waste hardboard. Laminate for trimming. 30 $1\frac{1}{2}$in (38mm) finishing nails. 50 $1\frac{1}{2}$in (38mm) No.8 Twinfast screws. 30 $\frac{3}{4}$in (19mm) No.6 screws. Laminate adhesive.

Bunk beds

Bunk beds

These bunk beds can be varied in length by replacing the side rails and inserting an additional strip to the plywood base.

This means that if the bunks are wide enough—at least 2ft 6in (762mm)—they can be made, initially, for children, and 'grow' with them. You will end up with adult bunk beds that are useful space-saving units for guests.

Construction details

The beds illustrated are made in teak, although any hardwood will do. Since teak is a very expensive wood you may want to exercise your option. If you are going to paint the wood, pick the most inexpensive hardwood you can find.

The unit basically consists of two rectangular frames, with plywood bases, mounted between four upright members as shown in Fig. 1. Both pairs of upright members or legs—the front and rear pairs—are joined across the bottoms by a 'skid' or runner as shown in Fig. 2. This not only makes the frame more rigid, but also enables the unit to be moved easily.

Various joints are used in the construction: a halving for the inside of the box frames; mortise-and-tenon for the top and bottom rails and the ladder; a single dovetail for the leg/skid joint; stub mortise-and-tenon for the head box joint; a mitered blind dovetail for the box corners at the bottom of the bunks; and a housing for the side tails.

The mitered blind dovetail (Fig. 10) is widely held to be one of the most difficult joints in woodwork, but this is not quite true. While it certainly takes a little patience and care to make, as long as the fit is reasonably snug, any imperfections will be covered by glue and hidden under the surface of the wood, unlike the open dovetail where any inaccuracies are instantly visible.

Upholstery details have been omitted here, because of the vast numbers of permutations possible. The mattresses can be purchased ready-made, or you can make them yourself from foam rubber or plastic with or without a fiber filling such as Dacron or another synthetic, and cover them in a suitable fabric.

The materials

A complete list of materials is given in the cutting list. Buy your lumber planed to size, but slightly overlength as there will be a certain amount of waste in finishing. Do not overdo this as it is only necessary to order each piece about 1in (25mm) longer than the final length intended.

Cutting and marking

With a pencil, lightly mark each piece of lumber so that you know where it fits in the unit—there is nothing more annoying than making a perfect joint, only to discover that it has been cut on the wrong piece of wood.

Carefully mark the length of each piece of lumber. When marking pieces of identical lumber, such as the four uprights, clamp or wedge the pieces together and mark them together so that each length will be consistent. Measure carefully several times before you mark and cut. Time spent on accurate marking is never wasted.

As most of the unit is held together by slot-type joints, it is advisable to cut these first. This will enable you to check your progress as you go along by fitting the various parts together in jig-saw fashion to make sure the dimensions are correct.

Mortise-and-tenon

Two types of mortise-and-tenon are used here. The cross bearers and ladder use the basic type as shown in Fig. 5; but the joints between the boxes and legs are mortise-and-tenons which have been haunched in the middle, as in Fig. 6.

Mark out all the tenons on the bearers first. Set the marking gauge to $\frac{3}{16}$in (5mm) (in order to mark out a $\frac{3}{8}$in [10mm] tenon), and run this along the ends of the pieces as in Fig. 5. Run it along all four edges and you will have marked out an area $\frac{3}{16}$in (5mm) deep (inward) all around. The tenon will stand $\frac{3}{4}$in (19mm) proud, so measure this distance down from the end of the lumber and mark a line right around. Now mark out the mortise outlines on the legs by laying the tenon over the leg at the correct place and direct marking. Tenon joints can be made with an electric drill and power circular saw or radial arm saw. See the Techniques section.

Next, mark out and cut the haunched mortise-and-tenon for the boxes and legs. Here again the tenon is $\frac{3}{8}$in (10mm) wide, but the shoulder is not the same width all around. It is cut $\frac{1}{2}$in (13mm) inward at the edges and $\frac{3}{16}$in (5mm) along the sides. The haunched portion is a 1in x $1\frac{7}{8}$in (25mm x 48mm) rectangle cut from the center of the leading edge as in Fig. 6. When you have finished these joints, put them together to check that they fit.

Dovetail

A single dovetail is cut out of both ends of each skid, which is fitted to the foot of each leg.

Mark out the dovetails on the skids as shown in Fig. 8. The shoulder is $\frac{1}{2}$in (13mm) wide and the slope is 1 in 7—but this is not critical. Cut the dovetails and use them as templates for transferring the outline to the legs. In this way each joint will be fitted invidually, ensuring a better fit. Clearly mark each pair of joints so that the correct match can be made easily, when assembling the unit later.

Stub mortise-and-tenon

The stub mortise-and-tenon joins the corners of the head box. The headboard is added when the bunks are finished. The stubs, as shown in Fig. 7, are $\frac{1}{2}$in x $\frac{3}{4}$in (13mm x 19mm) tenons. They are not cut away at the sides of the lumber and are only $\frac{1}{2}$in (13mm) deep. The top and bottom tenons are set $\frac{1}{2}$in (13mm) inside the edges of the boards and the mortises are $\frac{3}{4}$in (19mm) from the ends of the boards. Once again, temporarily fit the joints together to check the fit.

Mitered blind dovetail

This joint fits the corners of the foot box and the dimensions for making it are shown in Fig. 10.

First mark each board to what will be its finished length. Next, butt the box together and check with the try square that it is square. The small dovetails on one side (Fig. 10D) are called pins, and the opposite dovetails are tails. The pins are cut first, and this section is used as a template for marking the tails.

Using the diagram in Fig. 10A, mark out a rabbet $\frac{9}{16}$in (14mm) wide and $\frac{3}{16}$in (5mm) deep on the lumber. Cut the rabbet and you will have a section as shown in Fig. 10B. Now mark out the pins; these are at approximate $1\frac{1}{2}$in (38mm) intervals and $\frac{1}{2}$in (13mm) from

Fig 1. Side elevation of the bunk beds.

Fig 2. End elevation of the beds.

Fig 3. Construction details of the ends of the units. It is the rails which screw to the end boxes of the beds that give the units their ability to 'grow'.

Fig 4. Exploded view of the bunk bed unit. Only one end is shown.

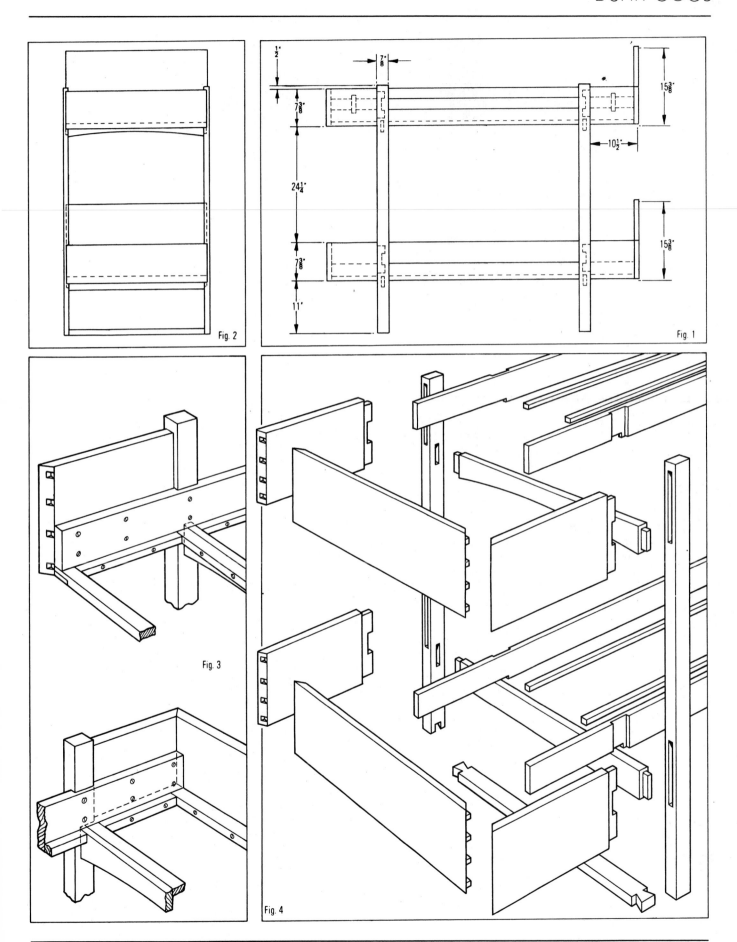

Fig. 2

Fig. 1

Fig. 3

Fig. 4

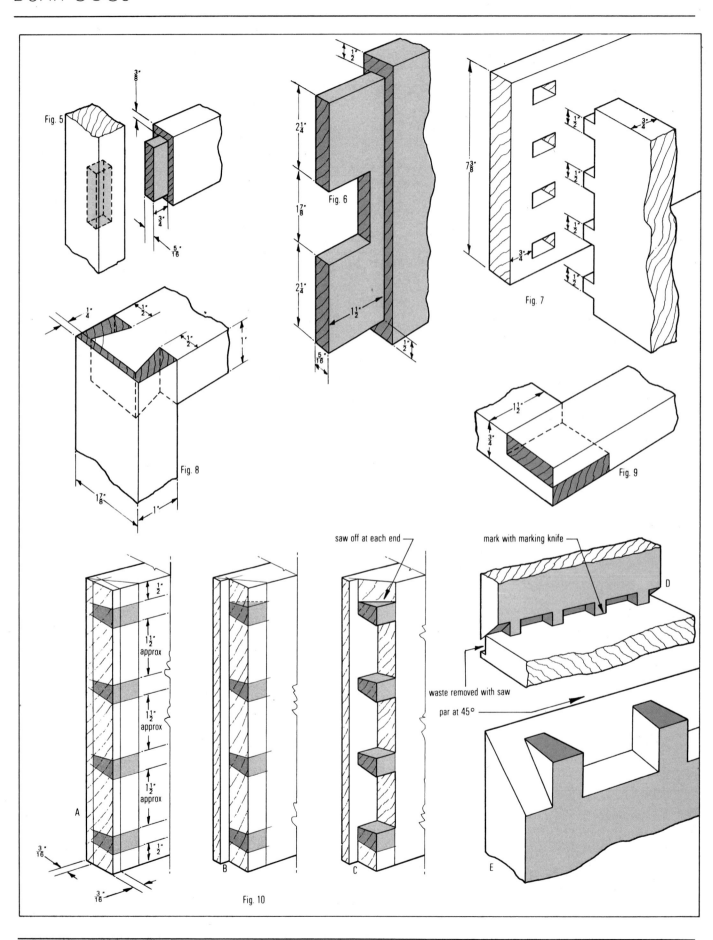

Fig. 5

Fig. 6

Fig. 7

Fig. 8

Fig. 9

saw off at each end

mark with marking knife

D

waste removed with saw

par at 45°

A

B

C

E

Fig. 10

each edge. With a crosscut saw and chisel cut the sloping recesses for the tails, leaving the pins as in Fig. 10C. Finish off by sloping the remaining lip of the rabbet with a paring chisel as in Fig. 10E.

Now cut an identical rabbet in the opposite board and, using the pins as a template, transfer the markings for the tails as in Fig. 10D. The tails can now be formed by cutting out the pin recesses.

When the joint is complete, mark each piece so that you can match it later.

Make the joint for the opposite corner of the foot box.

The joints for the corners of the inside box frames are of the simple halving variety shown in Fig. 9. Cross-lap joints (see Fig. 3) are used to join the legs to the side rails.

Assembling the units

The unit, from the assembly aspect, consists of two frames, each consisting of two legs, two struts or bearers, one skid, and two foot or head boxes—an example of the head frame is shown in Fig. 3. The two frames are stood upright, and joined by the side rails, which are screwed to the inside of the boxes and legs as shown in Fig. 4. An internal box framing, which is virtually battening around the inside, is then screwed around to support the edges of the plywood panel which is the base for the mattress.

First, glue, fit and clamp the head frame as shown in Fig. 4. Lay the assembly down on a flat surface for this, and constantly check all angles for squareness. It is best to complete all the head frame at the same time—if you make the boxes separately, the tenons might not fit into the mortise slots when the glue is dry. Repeat this for the foot frame.

When the adhesive has thoroughly set on both frames, place them on their sides so that the open ends of the boxes are facing the middle, lay the rails along the inside and mark out the joints to fit over the inside of the legs as in Fig. 3. Cut the joints and screw the side rails in position to the legs and boxes which are on the floor. Carefully lift the frame over so that the opposite side is on the floor, and mark, cut and screw the remaining side rails in position.

Fit and screw the box frame battening around the inside of the bunks as in Fig. 3. Then, fit and screw the plywood mattress bases to the top of the battening.

The guard rails for the top bunk can be fitted in the same way as the rails. This is recommended for children because they cannot remove the guard rails. However, there are many fittings which provide for easy removal of the guard rails and this might be an advantage for adults. One such method consists of brass rods sunk and glued into the bottom edge of the guard rail and fitted into holes drilled in the top edge of the side rails.

The headboards are simply cut away at the lower outside edges to provide recesses for the protruding corners of the sides of the head boxes, and then screwed to the top of the head box.

The ladder

This consists of five rungs fitted into the sides by means of mortise-and-tenon joints, and is simple to construct. Round off the tops of the rungs slightly so that they are easier on bare feet.

The ladder hooks can be made from mild steel strip, but there are many types available on the market that you might prefer.

Finishing

Sand down all surfaces using an orbital finishing sander with a 100 grit sandpaper, wipe with a cloth dampened with turpentine to remove the wood dust, and varnish with a clear polyurethane diluted 50/50 with turpentine. When dry, rub down with a grade '0' steel wool, wipe it down and finish with polyurethane.

Fig 5. The basic stopped mortise and tenon used in the construction of the bunk beds.

Fig 6. The end boxes joint to the uprights using a mortise and tenon haunched in the middle.

Fig 7. The head boxes are jointed using stub mortise and tenons.

Fig 8. The stopped dovetail which joins the skids to the uprights.

Fig 9. The half lap used to join the inside corners of the box frame.

Figs 10 A–E. The method of cutting the mitered blind dovetails

Cutting list

Solid wood	standard	metric
Teak		
4 legs	1 x 1⅞ x 50½	25 x 47 x 1283
2 skids	1 x 1⅞ x 32	25 x 47 x 813
4 cross bearers	¾ x 3 x 32	19 x 76 x 813
4 foot boxes	¾ x 7⅜ x 12	19 x 187 x 305
2 foot boxes	¾ x 7⅜ x 31¾	19 x 187 x 806
4 head boxes	¾ x 7⅜ x 12	19 x 187 x 305
2 head boxes	¾ x 7⅜ x 31¾	19 x 187 x 806
4 side rails	¾ x 3½ x 31¾	19 x 89 x 806
2 safety rails	¾ x 1¾ x 31¾	19 x 45 x 806
2 ladder sides	¾ x 1½ x 49½	19 x 38 x 1257
5 ladder rungs	¾ x 1½ x 12½	19 x 38 x 318
Hardwood		
4 internal battening	¾ x 1½ x 44	19 x 38 x 1118
8 internal battening	¾ x 1½ x 10	19 x 38 x 254
8 internal battening	¾ x 1½ x 32	19 x 38 x 813
Firwood		
2 mattress bases	½ x 29¾ x 64	13 x 756 x 1626
2 headboards teak or flakeboard	¾ x 15⅝ x 32½	19 x 391 x 826

Loft bed

A proportion of today's houses consist of buildings 40 years or more old. By modern standards the ceilings in such homes are high, and the top part of each room is just wasted space. A loft bed enables you to use some of this otherwise wasted area —and add interest as well as utility to your sleeping space.

This particular structure has been designed to provide a platform, some 7ft (2.1m) high, covering a sufficient area for sleeping space for three adults. A large room is required for this, but you can alter the dimensions to suit your room. For example, in a smaller room you could build the storage/desk unit with single bed above, as shown in Fig. 1. In a small apartment this could be the sleeping accommodation for the occupant, while in a larger home, it might be used for the occasional visitor.

The boarded-in area at the base could provide even more sleeping space—in which case this particular section would be nothing more than a large pair of bunks beds—but here it is used as a table of sorts, with storage space underneath that is reached by doors at one side. Alternatively, you could use this as a desk in which case one of the doors should be left out to provide a recess for your knees.

Construction

The main supports consist of 1⅛in x 3½in (28mm x 89mm) members. The two rectangular frames shown in Fig. 4 are of 1⅛in x 2½in (28mm x 64mm) lumber. The sides of this unit may be boarded in with the material of your choice. Tongue-and-groove boarding may be used if it is available commercially. Otherwise, for those with access to a router, it may be made from 4¼in (108mm) wide planking. The plank must be rabbeted on both sides of one edge, to one-third of the thickness of the wood. The other edge having a groove cut to a depth of ¼in (6mm) and a width of one-third of its thickness to accept the tongue. Finally, it is good practice to round the edges of the tongue to make fitting easier. Two simpler alternatives would be to use 4in (102mm) wide planking of ½in (13mm) thick plywood in which case the dimensions must be taken from the completed frame.

The top, or platform, has a framing of 1⅛in x 3½in (28mm x 89mm) planking bolted to the vertical supports. The method of doing this is shown quite clearly in Fig. 2 which is a 'top floor' plan.

The floor of the platform consists of 1in (25mm) thick tongue-and-groove boarding laid over an inside rim of 1in x 1in (25mm x 25mm) battening screwed around the planking as shown in Fig. 6.

Tongue-and-groove boarding, when used for flooring, is secured across the width—that is, the narrowest length—of each frame. If laid across the length it would give too much in the middle. For this reason, where a frame is wider than 3ft (914mm), the underside of the flooring should be reinforced every 3ft (914 mm) of the frame, with 1in x 2in (25mm x 50mm) battening. Once again the platform may be planked or filled in with plywood.

A major safety point to note is that the combined unit, as shown in Fig. 2 is a self-supporting structure. But if you wish to build only the main unit as shown in Fig. 1, then it should be built against a solid wall the main supports on one side bolted or otherwise secured to the wall. This is necessary because the weight of an adult positioned at one top edge—as happens when climbing on and off— could cause the structure to sway and possibly topple over if it were not firmly fixed to the wall.

The main frame

This is the section with a boarded-in base shown in Fig. 1.

First mark out the floor area.

Stand two of the uprights against a wall, measure a line 5in (127mm) downward from the top of each, then drill holes and secure the platform side across the uprights with one top bolt at each end. The top of the platform side should be level with the marked line.

Now fit the platform end, with one top bolt, to one of the uprights as shown. The platform end will be 'hanging' from this one bolt. Stand another support in position so that you can swing the platform end up and butt snugly against it pulling the other two supports upright. Drill and bolt the end in position.

You now have a 'dog-leg' consisting of three supports held upright by two horizontal or cross-members. Because the cross-members are only held by a single bolt at each end, it is possible to hold a plumb line by the side of each support and adjust the base until each member is absolutely vertical. When you are sure that every support is vertical, use the same procedure you used for the platform end to complete the remaining

Loft bed

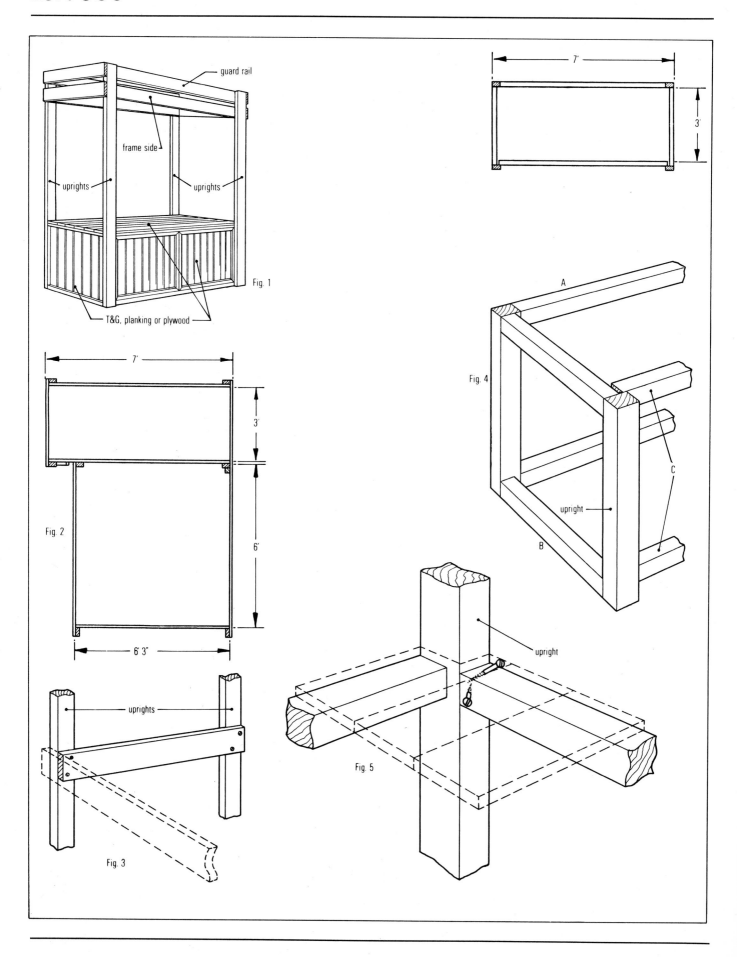

guard rail

frame side

uprights

uprights

T&G, planking or plywood

Fig. 1

7'

3'

Fig. 2

7'

3'

6'

6' 3"

Fig. 3

uprights

Fig. 4

A

B

C

upright

Fig. 5

upright

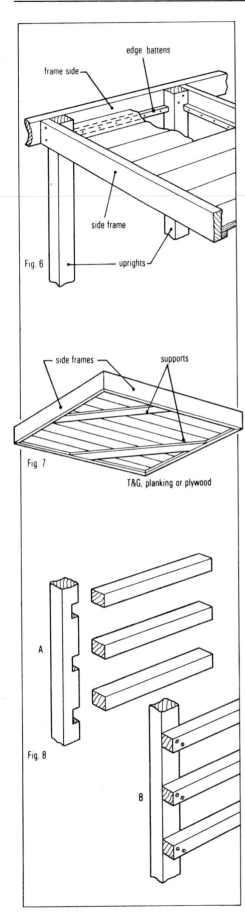

edge battens

frame side

side frame

Fig. 6

uprights

side frames

supports

Fig. 7

T&G, planking or plywood

A

Fig. 8

B

side, end, and support. When this has been done, check the whole assembly for squareness and drill and fit the remaining eight bolts.

Next make the framing for the base with 1⅛in × 2½in (28mm × 63mm) lumber. The ends of each member are first nailed to a support with finishing nails toed through the framing into the support, then screw holes are toe drilled through each end on the top, bottom and rear faces, into the support member. This applies to members A and B as shown in Fig. 4, but not to member C which is screwed directly to the inside of an upright as shown in Fig. 4. Member A must also be shortened to 6ft 5in (1955mm)—see Fig. 4.

Fix members A and B at the bottom of the unit first. Then fix the upper members A and B at a height of 2ft 3in (885 mm). Figs 1, 4 and 5 show how members A, B, and C butt or join to the uprights.

When the frame is complete, fit the ends in place. These will have to be notched around pieces A and C. Glue and screw the plywood top and bottom with 1½in (38mm) No.8 flathead steel screws spaced about 6in (152mm) apart.

Next, fit and secure one side panel. The other side panel must be cut to create two sliding doors to ride in aluminum sliding door hardware. If tongue-and-groove boarding has been used, this must be braced with solid lumber. Another option is to leave all or part of this side open so that it can be used as a desk.

The base is finished by installing the top. This piece must be fitted around the four uprights. The edges of the top will be visible so cut and sand them carefully. Of course, there are a variety of methods you can use to hide them.

To complete the platform, screw a frame of 1in × 1in (25mm × 25mm) battening around the bottom of the inside faces of the planking you fixed first of all. The battening should be secured with flathead No.8 wood screws at 3in (76mm) intervals.

When the frame is finished, lay the panels. Glue and screw them in place with 1in (25mm) No.8 flathead wood screws about 6in (152mm) apart.

It only remains to fit the side and end guardrails around the top. This is 1⅛in × 2½in (28mm × 63mm) lumber and is bolted to the supports in the same way as the side and the end immediately below.

Fig 1. The basic unit. If this is to be used on its own, the uprights must be secured to a wall, otherwise the weight of a person climbing into it may cause it to topple over.

Fig 2. Plan view of the unit with extension. In this form the loft bed is free standing.

Fig 3. Method of fitting the platform rails to the uprights.

Fig 4. The cabinet construction with plan view of the same. Note the way in which the front horizontal fits inside the uprights.

Fig 5. The top of the cabinet unit must be notched around the uprights as illustrated here. Note the method of screwing the horizontals to the uprights.

Fig 6. Method of constructing the side unit.

Fig 7. The base of the side unit requires support whichever method of construction is used.

Figs 8 A–B. The two alternative methods of constructing the ladder.

Loft bed

The side unit

This is really only an extension of the main frame. Here it has been made wide enough to accommodate two sleeping adults, but the dimensions can easily be altered.

The method of construction is identical to that of the main unit. As shown in Fig. 2, two supports are bolted to one side of the main unit, and these provide fixing points at one end for the side of the unit, with the two remaining supports placed at the far end as shown in Fig. 6. These supports should be braced to the floor at their lower end.

When a side unit is added to a main unit, as shown here, the guardrail separating the two is omitted or removed to allow easier access to both sides.

If you make a side unit as large as the one shown here, then you will have to fit underfloor supports underneath, so that it will not sag in the middle, and to be a butt block between the two pieces. Use two pieces of 1in x 2in (25mm x 51mm) lumber fixed diagonally (Fig. 7). The butt joints with the frame must be cut and fixed accurately to provide the required support and should be glued and screwed from outside the frame.

The ladder

This ladder has rungs that are housed into the side supports as shown in Fig. 8A. To do this, cut all rungs to an exact length and directly mark, cut and fit so that each rung houses perfectly. The joints are secured with glue and screws skewed through the rungs into the main side supports. Alternatively, the screws can be simply inserted through the main supports into the end grain of the rungs.

A simpler version is shown in Fig. 8B. Here the rungs are laid directly over the side-rails and secured at each joint with glue and two screws.

Whichever method you use, when you are finished round off the tops of the rungs with a spokeshave in order to make climbing up and down the ladder easier – and more comfortable on your feet.

With a loft bed you have more living space and the concept of a home-in-a-room can become a reality. And even if you are already reasonably well-off for space, a loft bed could release a bedroom for other purposes. But a word of warning. If you make this loft bed for children, they'll like it so much you will have a job getting them off it!

Cutting list

Solid wood	standard	metric
Main unit		
4 uprights	$1\frac{1}{8}$ x $3\frac{1}{2}$ x 8ft	28 x 89 x 2.4m
4 lower frame sides	$1\frac{1}{8}$ x $2\frac{1}{2}$ x 6ft $9\frac{3}{4}$in	28 x 64 x 2.1m
4 lower frame ends	$1\frac{1}{8}$ x $2\frac{1}{2}$ x 3ft	28 x 64 x 914
Lower frame boarding		
21 tops (T&G)	$\frac{1}{2}$ x 4 x 3ft	13 x 102 x 914
42 sides (T&G)	$\frac{1}{2}$ x 4 x 2ft	13 x 102 x 610
18 ends (T&G)	$\frac{1}{2}$ x 4 x 2ft	13 x 102 x 610
Platform materials		
2 sides	$1\frac{1}{8}$ x $3\frac{1}{2}$ x 7ft	28 x 89 x 2.1m
2 sides	$1\frac{1}{8}$ x $2\frac{1}{2}$ x 7ft	28 x 64 x 2.1m
2 ends	$1\frac{1}{8}$ x $3\frac{1}{2}$ x 3ft $2\frac{1}{4}$in	28 x 89 x 972
2 ends	$1\frac{1}{8}$ x $2\frac{1}{2}$ x 3ft $\frac{1}{4}$in	28 x 64 x 921
2 edge support battens	1 x 1 x 7ft	25 x 25 x 2.1m
2 edge supports	1 x 1 x 3ft	25 x 25 x 914
21 flooring (T&G)	1 x 4 x 3ft	25 x 102 x 914
Side unit		
4 uprights	$1\frac{1}{8}$ x $3\frac{1}{2}$ x 8ft	28 x 89 x 2.4m
Platform materials		
2 sides	$1\frac{1}{8}$ x $3\frac{1}{2}$ x 6ft	28 x 89 x 1.8m
2 sides	$1\frac{1}{8}$ x $2\frac{1}{2}$ x 6ft	28 x 64 x 1.8m
1 end	$1\frac{1}{8}$ x $3\frac{1}{2}$ x 6ft 3in	28 x 89 x 1.9m
1 end	$1\frac{1}{8}$ x $2\frac{1}{2}$ x 6ft 3in	28 x 64 x 1.9m
2 edge support battens	1 x 1 x 6ft 3in	25 x 25 x 1.9m
19 flooring (T&G)	1 x 4 x 6ft	25 x 102 x 1.8m
2 underfloor supports	1 x 2 x 7ft	25 x 50 x 2.1m
The ladder		
2 sides	$1\frac{1}{8}$ x $2\frac{1}{2}$ x 7ft	28 x 64 x 2.1m
10 rungs	$1\frac{1}{8}$ x $2\frac{1}{2}$ x 12	28 x 64 x 306mm

You will also require:

2 brass 3in (76mm) angle brackets. At least twenty-four $3\frac{1}{2}$in (89mm) long $\frac{1}{4}$in (6mm) bolts with nuts. Wood screws, No. 8 $1\frac{1}{2}$in (38mm) and 3in (76mm). Finishing nails $1\frac{1}{2}$in (38mm). Wood adhesive.

Note: If T&G boarding is being made, $\frac{1}{4}$in (6mm) must be added to the width dimension to allow for the tongues. Dimensions for ordinary planking will remain the same while the dimensions for a ply version may be taken from the finished frame.

Closet

This closet is both simple to construct and a great space-saver.

This closet has been built into a corner, but it could equally well be built against a wall or into an alcove.

General construction details

The construction outline, with exploded views of the joints used, is shown in Fig. 1. A detailed parts list is not given here because it is certain that the dimensions of your house are not identical. When making built-in units such as this you

have to design according to the room you are fitting. For instance, this particular unit is 7ft 5in (2.3m) at the back, because of the slope of the ceiling.

Main supports consist of three 2in x 2in (51mm × 51mm) vertical members, but a closet built against a wall would require four supports, or only the two front ones if built into an alcove. Lumber of the same dimensions is used for the front bottom and top cross-members, and for one cross-member in between these. The middle rail, 1½in x 2in (39mm × 51mm), acts as the main

support for the lower sliding doors set between two runners. A section view of the runners for both sets of doors is shown in Fig. 2. Runners and fittings for both upper and lower doors are commercially available. There are many different types on the market and the ones shown here are just two examples. Your local retailer will show you what is available.

The fittings of the lower runners are hidden from view by ½in x 4in (13mm x 102mm) wood valance as shown in Fig. 2. This valance is mortised into the vertical supports at each end and also acts as a

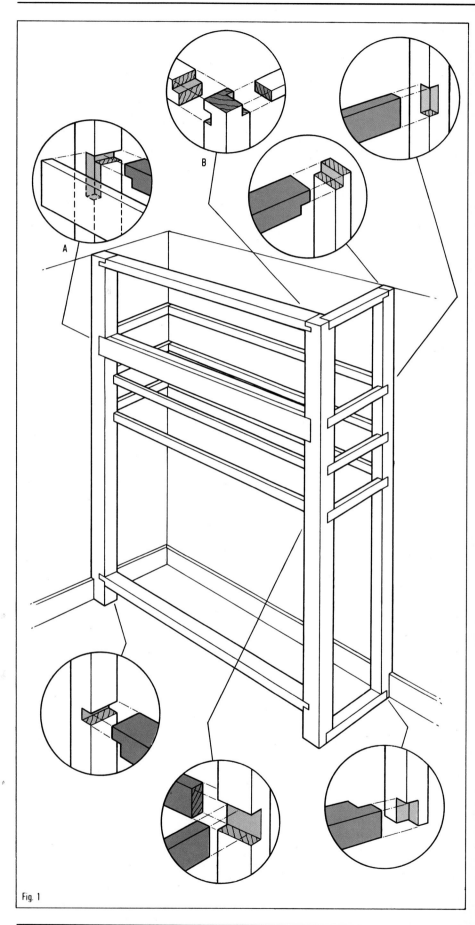

Fig. 1

Fig 1. Construction details for a closet built into the corner of a room. The third upright is unnecessary in this case. The unit is simply built—all joints being of the ordinary housing type.

strengthening support for the middle rail.

The frames for the three shelves consist of 1¼in x 2in (31mm x 51mm) pine. The closet is a floor-to-ceiling structure and is secured to the walls at two points as shown in Fig. 1. If you are building in a corner or alcove, it is not necessary to screw the top and bottom members to the floor and ceiling, but this will have to be done in the case of a unit with two free-standing sides as in Fig. 4.

The receiving wall

It is essential for the supports on each side of the doors to be absolutely vertical. The rear supports, if any, can be screwed to a wall that is out of true, and the joints adjusted to fit. But if the front supports are not in plumb, the doors will not close flush with the sides of the closet.

The easiest way of fixing a lumber support vertically to a 'leaning' wall is to stand the support so that it is touching the wall at one end.

Use a plumb line or level to ensure that the lumber is vertical. Then measure the space between the lumber and wall at the 'open' end. Put the lumber aside and screw a block of wood, the same thickness as the space and as wide as the lumber, to the wall. The lumber can then be screwed to the block at one end and to the wall at the opposite end, and the wedge-shaped space between it and the wall filled with plaster. When the plaster has set, drill holes through the support firmly in place. If required, use toggle bolts. This method is the one to use if the wall is an inch, or less, out of true.

If your wall is more than an inch out over the length of the wood support, it would be better to cut a wedge-shaped fillet to shape as shown in Fig. 5. A piece of lumber, as wide as the support but thicker is placed vertically against the wall, with a 1in (25mm) piece of batten at the 'narrow' end (the batten can be dispensed with if a baseboard is there). Ensure that the support is vertical and place a block into the wide end as shown in Fig. 5. Now measure the gap between the edge of the lumber nearest the wall and the wall itself at regular intervals and lightly mark the lumber at these points. Lay the lumber on the ground and mark

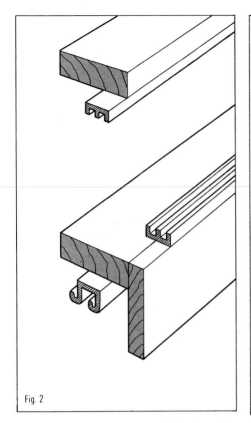

Fig. 2

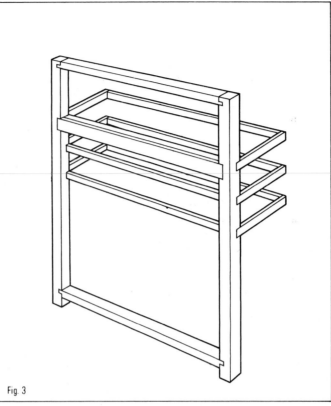

Fig. 3

Fig 2. The type of door runners used is a matter of personal choice.

Fig 3. If the closet is to be built into an alcove, the rear uprights may be dispensed with altogether. The basic construction remains the same in all cases.

off each measurement along the length. Draw a line along the ends of these as shown by the dotted line in Fig. 5, and cut the lumber along the line. This is easily done with a table saw. See the chapter on power sawing in the Techniques section.

When the cut-out portion that has the markings is turned around, it should fit the incline of the wall perfectly. If there is a baseboard at floor level, the bottom of the fillet will have to be sawn off level with this so that the fillet can be screwed directly to the wall.

Building the main frame
With chalk, mark out the outline of the closet base on the floor. This closet will only require two lines—side and front. Make sure that the area is in square by measuring the diagonals and checking that both measurements are identical. Using a plumb line, transfer the marks to the ceiling.

Measure the height of the uprights—they should all be about the same length, so cut them to the shortest measurement. Put the two front uprights together and mark the positions of the rails. Cut the housings and then cut the rails to length. Mark the shoulders of the joints by placing the rails together and squaring the lines across all three rails. This will

ensure that the uprights will be parallel.

Cut the joints and fit the frame together dry, to check that all the dimensions are correct. Then, glue the joints, fit them together and nail them to hold them while the glue sets.

Before the glue sets, measure the diagonals to ensure that they are the same length and that the frame is square. Then tack a temporary batten from the middle of the top rail across the middle rail and to about the middle of the upright. This will hold the frame square until it is fixed in place.

Mark the end joints, putting the third upright against the frame flush at the top and bottom, and squaring the lines across the two timbers. Then cut out these housings. Cut the end rails and glue and pin them into the housing of the third or back upright. The front frame can then be lifted up and the end section glued and nailed to it. The wood valance can be either fixed now or later when the doors and tracks have been fitted. The latter may be the best as it will give more freedom to fit the door.

Now you can put the whole framework into position and mark the walls so that they can be drilled and plugged. Three screws to each upright should be sufficient. The front frame is square, so if you level the bottom rail, the sides will be plumb.

Use a little packing if necessary. If the upright does not fit the wall it can either be cut into the plaster or trimmed with a plane to fit. If the gap is too large to plane the side to fit, you will have to pack the upright off the wall when you drive the screws home and cover the resulting space with a small quadrant moulding. Measure the end frame and the distance from the back wall to the front frame at the other end to ensure that the front frame is parallel to the back wall.

Measure and cut the shelf supporting battens and plug and screw them to the wall finding their position by leveling from the the top of the end rails.

Shelves and side
Each shelf top is marked out and cut individually. You can do this by taking several length and width measurements for each shelf space and transferring them to the panel of hardboard or plywood from which they will be cut. However, it is much better to make paper templates for each shelf and use these for marking out. A large template can be made of lengths of newspaper or wallpaper joined with cellulose tape.

If you are using hardboard for the shelving, then this must be nailed all around the shelf supports with brads. If you are using plywood, it is a good idea

Closet

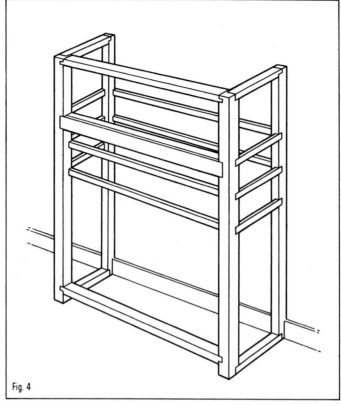

Fig 4. A free-standing unit requires a different approach. Four uprights are necessary to provide adequate support for the structure.

Fig 5. Walls are rarely absolutely true and vertical. The uprights need to be trimmed accordingly, using a level to make sure that the lumber is plumb.

Fig. 4

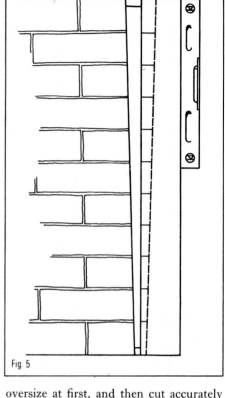

Fig. 5

just to rest the shelf panels on the shelf supports. This way you can remove them easily for cleaning, or if you want to alter the internal layout, or perhaps to repair the sliding door runners should it become necessary.

Fitting the hardboard or plywood end will need a paper template; but if you cannot make one, cut the plywood to the height of the framework and a little wider than the end framework. Put it in position with one side against the wall and the other parallel with the face of the front frame. If when the front edges are parallel the plywood does not fit the wall you can scribe it by taking a small piece of wood or ply about $\frac{1}{4}$in (6mm) wider than the widest part of the gap to make a guide. Slide this piece of wood down the wall making a mark on the plywood side with a pencil held against the outer edge of the wooden guide. The result should be a line that is parallel with, and as uneven as, the wall. Cut the ply along this line and prop it in place again; this time measure how much waste should be taken off so that the panel fits flush with the front frame.

Nail the panel to the framework with finishing nails with their heads punched below the surface and the holes stopped with suitable filler.

A length of skirting board will cover

the joint at the bottom of the end panel and this board can be continued around the front of the frame to meet the skirting on the other wall giving a more professional finished look.

The doors

Sliding doors of the type used for the lower part of this closet are, in fact, much easier to fit than conventional hinged doors. They are merely hung on the top runners and dropped into a slotted plate below. The runners shown in Fig. 2, are fairly typical, but they do vary in fitting. Pay particular attention to the instructions that come with your runners. It is in this area of construction that most handymen make a mistake. The key is to take your time and to do the job carefully. Should you have any questions you can usually get any information you require simply by consulting either the sales-person or the manufacturer.

The top sliding doors are more simple and consist of two lengths of grooved wood into which sheets of hardboard are fitted as shown in Fig. 2. The strips or runners are nailed in position at the top and bottom of the upper compartment, and flush with the front of the closet. The top grooves must be twice the depth of the bottom grooves. The two doors are cut from a sheet of hardboard, slightly

oversize at first, and then cut accurately using direct marking. The doors must be fractionally higher than the distance between the runners, so that each door can be eased into the top recess, straightened, then dropped into the bottom recess. The priming, undercoat and topcoats of paint should be applied to these doors before they are finally fitted.

For both sets of doors, ordinary door handles would, of course, be impossible to use because they protrude and would stop the doors sliding. But there are many recessed pulls for sliding doors on the market that your local retailer will show you.

Finishing

Smooth down any bumps or rough patches in the woodwork with a plane. Using a spatula and a cellulose filler, fill in any visible holes or cracks. Apply the filler generously, so that it dries higher than the surface of the hole or crack. Then smooth the filler down level with the surface with fine sandpaper. Rub the entire surface of the closet down to a smooth finish and apply a layer of undercoat. If necessary, rub down again, then finish with a coat of undercoat, followed by two coats of paint, allowing plenty of drying time between applications.

Built~in dressing table

The simple design of this dressing table is adaptable to any size room. The unit is functional, allowing ample shelf and storage space.

Materials

The dressing table is made mostly from $\frac{3}{4}$in (19mm) thick plywood, but any softwood that can be sanded to a smooth finish is suitable. The dressing table top, the shelf panel and the bottom and top panels of the upper component are made from plywood; so are the bottoms of the drawers. The drawers run on strips of hardwood fixed under their outside bottom edges.

The only joints used in the construction are simple butt joints, which are held together with glue and finishing nails. The central shelf is held up by standard plastic shelf supports fixed to the vertical dividing pieces. The wall at the back of the shelf area is covered with plastic laminate, and the area above the dressing

table with mirrors. These are fixed with screws positioned so that they are not conspicuous when the unit is installed. In this way, you can save the trouble of making a back for the unit itself.

The sides of the unit

The dimensions of the unit's sides are shown in Fig. 2. These pieces are fixed to the wall first, and the horizontal pieces of the unit fitted between them.

First, cut the ends and vertical dividers to the length and width required and square their ends. Lay them on a level surface and mark the position of the top piece, the facing strip of the middle shelf and the dressing-table top. The facing strip of the shelf is let into the front edge of the side for a neat appearance.

From the top of the ends and dividers measure downward a distance of 10in (254mm). This indicates the top of the housing. The housing is $\frac{3}{4}$in (19mm) deep and $2\frac{3}{4}$in (70mm) long. Mark the depth of

Above: The completed dressing table in position. It is both functional and attractive; and the dimensions can be varied to suit individual requirements.

Built~in dressing table

Opposite page: This dressing table has been made mainly from plywood, but any softwood that can be sanded down to a smooth finish is suitable.

Fig 1. Construction details of the dressing table. The supports for the table top also provide a guide for the drawers.

the housing with a marking gauge and use a try-square to mark the ends of the housing. Cut the ends with a tenon saw and chisel the housings out to the correct depth.

The next step is to fix the ends and dividers to the walls. On the walls, mark where the bottom ends of the sides will come. These points are 23in (584mm) from the floor on each wall. This will make the dressing table top $27\frac{7}{8}$in (708 mm) high—about right for a person of average height. Raise or lower it if you like. Draw a string from mark to mark and check it for level.

Now fasten the side pieces to the wall. The original of this unit was fixed between two end walls, so the sides did not

need to be screwed on through the back edge. They were just screwed in place through their flat surfaces, with the screws arranged to fall behind the horizontal pieces so as not to show. If your unit is to be free-standing, you should mount the sides on pieces of aluminum angle strip.

The top of the unit

A top for the unit is made out of $\frac{5}{8}$in (16mm) plywood fixed to a wood framework. Cut two lengths 1in x $2\frac{3}{4}$in (25mm x 70mm) to the full length of the unit. Cut four pieces of $\frac{3}{4}$in x $1\frac{1}{2}$in (19mm x 38mm) $6\frac{3}{4}$in (171mm) long and two the same length out of 1in x $2\frac{3}{4}$in (25mm x 70mm). Mark the positions of the dividers onto

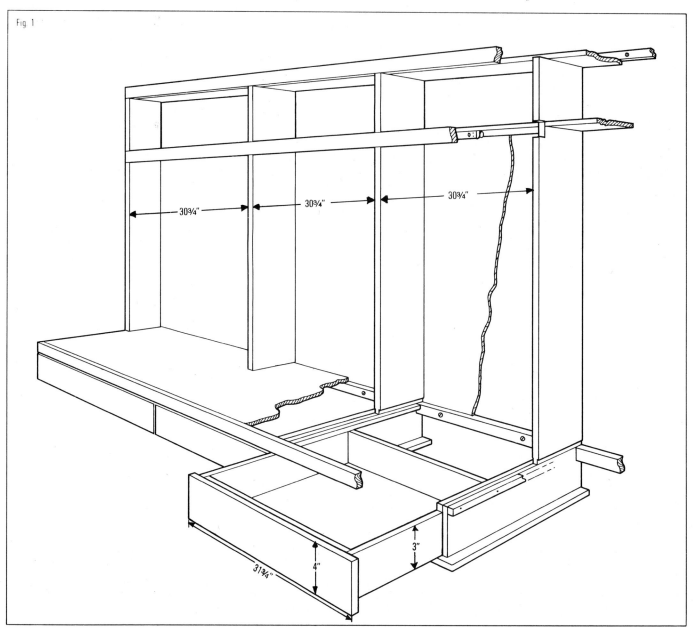

Fig. 1

30¾″   30¾″   30¾″

3″

4″

31¾″

the plywood top piece. If the plywood top has to be made of more than one piece of wood, try to make the joint over the top of one of the dividers. The plywood top will be 6¾in (171mm) wide, and is set out to take the four intermediate pieces of lumber and the two ends.

Glue and nail the intermediate members into place flush with either side of the plywood panels. Then fix the ends and the front and back members by gluing and nailing them to the ends of the intermediate pieces and the edge of the ply. The top section can then be lifted up and fixed firmly by first screwing it to the tops of the dividers and next by plugging and screwing this section to the wall, making sure it is firmly joined.

The table top

The table top is made from ¾in (19mm) plywood. At the front, a strip of ¾in × 1¾in (19mm × 44mm) lumber is glued and nailed to it. The back of the table top, which butts against the wall, is supported by, and fitted to, a strip of ¾in × 2in (19mm × 51mm) softwood plugged and screwed to the wall. The plywood top of the table is supported at the dividers by 1in × 1in (25mm × 25mm) softwood battens.

The plywood top has deep slots cut into it to enable the vertical dividing pieces to fit into it. You will not be able to get a single board of the length required for the top, so use two or more boards. Make sure that the ends of the board that are to butt are perfectly square. The joints will be less conspicuous if they occur parallel to, and in line with, the slots.

The table top an now be fitted to the wall.

The shelf

The central shelf is made up from ⅝in (16mm) plywood panels which extend between pairs of vertical dividing pieces. They are supported where they meet these pieces by small commercial nylon shelf supports screwed in place. The front facing strip, which is housed into the dividing pieces, should be fitted with these nylon supports too so that it supports the front edge of the shelf. The supports should go ⅝in (16mm) below the upper edge of the facing.

Cut the facing strip to length and glue and nail it in place. Cut the plywood shelf panels to size. When they are fixed, their front edge should be flush with the rear surface of the facing strip. Fix the nylon shelf supports and put the shelves in place on them.

The drawers

The unit has four drawers made from ¾in (19mm) lumber with ⅝in (16mm) battens. The four sides of the drawer are butted, glued and screwed together. A front panel is glued to the front side of each drawer and screwed on from the inside. These panels protrude below the bottom edge of the front side and beyond the end of the front side.

The dimension of the drawer components are shown in Fig. 1. Cut the four sides to length and square their ends. Drill screw holes in the pieces, lay them on a level surface and apply adhesive to the ends of the two short sides. Screw

Built-in dressing table

the pieces together. Check that the construction is square and lightly bar-clamp it (or use string twisted tight with pegs if you don't have any bar-clamps).

The drawer facing protrudes below the bottom edge of the front side of the drawer by 1in (25mm). The front pieces are 31¾in (706mm) long, so that in the finished unit the ends of one drawer unit almost butt against that of the adjacent drawer unit.

Lay each front piece on a flat surface with its inner face uppermost. Mark from each end the distance it protrudes beyond the front side of the drawer. Lay the partially finished drawer between these two points and draw around this outline. Apply adhesive to the front piece between the pencil lines and on the front side of the drawer. Finally screw the pieces together from the inside and clamp them firmly together.

The drawer runners

The drawer runners are glued and screwed to the underside of the dividers. Cut hardwood strips ¾in x 2½in (19mm x 63mm) as long as the L-shaped projection of the table top supports. Then glue and screw them into place, positioning the strips centrally so that they will carry a drawer on each side of the divider.

Finishing the unit

Sand all the surfaces smooth. Knock any finishing nail heads that are conspicuous below the surface of the wood. Fill the holes with wood filler. Apply several coats of polyurethane varnish to the unit, sanding down carefully between individual coats.

The mirror

The mirror is fixed in place with mirror screws positioned 1in (25mm) from the corners of the mirrors. Your glass dealer should supply mirrors cut to size and ready-drilled. The plastic laminate that backs the shelf can simply be glued to the wall.

Fig 2. The correct method of laying out the lumber panel from which the vertical partitions of the dressing table unit are going to be cut.

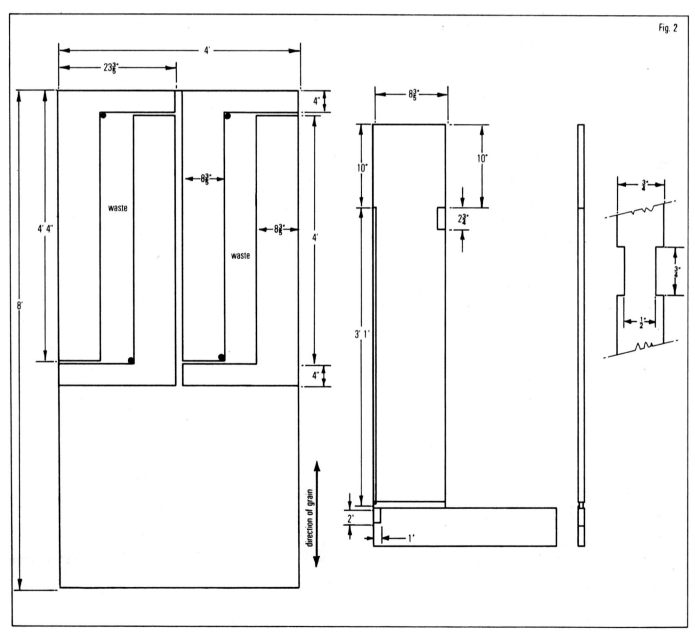

Four-poster bed

Four-poster bed

Fig 1. The panels for the four-poster bed are all cut from three sheets of plywood. The layout for the first of these is shown here.

Figs 2-3. The layout for the other two sheets. An allowance must be made for the sawcuts and great care must be taken when cutting these to ensure that they are straight and true.

Fig 4. Detail of the ends of the panels where they meet the uprights at the foot of the bed.

Fig 5. The dimensions for the headboard.

Fig 6. View of the rear of the headboard showing the position of the batten.

Fig 7. View of the front of the headboard showing the position of the lower batten.

Fig 8. Lower side panel of the bed and position of the support batten.

Fig 9. Lower end panel of the bed showing position of the support batten.

Fig 10. The top panels of the bed showing positions of the battens.

The four-poster bed is the aristocrat of bedroom furniture. It has the flavor of history and an element of prestige. Unfortunately, four-poster beds are not sold by most furniture shops. Where they are, only traditional designs are available and these are expensive. Instructions are given for making a four-poster of modern design that does not cost the moon. If you prefer, you could provide a traditional touch by adding curtains.

Construction

This bed consists of four upright 3in × 3in (76mm × 76mm) deal posts—you could use another wood if you wish—supported and kept vertical by eight panels, four at the base and four at the top.

The panels are of firwood, and at the base butting to the floor, there are two side panels and one footboard. The panel at the head end is raised to provide support and to act as a head board.

The four panels or fascia running around the tops of the uprights provide rigidity for this part of the structure, but are also used for decoration. They can be painted, covered with fabric, or fitted with a curtain rod.

Firwood rarely has an attractive surface, and the end-grain or cut edges look unsightly. For this bed, therefore, the firwood panels have been covered with felt, but you could use some other material, or paint. If you prefer a natural wood finish, then the firwood would have to be replaced by hardwood, veneered firwood or with ¾in (19mm) solid wood planking, but this would be much more expensive. If you do decide to do the latter, then each base panel will be replaced with four solid planks as wide as the panels they replace. The solid planks can be glued and clamped together to form solid panels, or left as individual planks. The main advantage of the former method is that it prevents the planks from buckling out of line.

Support battens are fixed all around the upper inside edges of the side panels. This provides support for the slats, which in turn makes the frame more rigid and provides the support area for the mattress. These slats are part of the structural design, and should not be replaced with something less rigid.

Materials

The uprights are of deal. Any other suitable wood can be used, but in any case the lumber for the uprights should be ordered as perfect as possible because in this design, the uprights are the only part of the woodwork that is visible. The rest of the solid lumber can be of a common pine. This consists of the slat supports and edge battens, which are of 2in × 2in (51mm × 51mm) members and the slats, which are 1in × 6in (25mm × 152mm).

All the panels are cut from ½in (13mm) firwood, but if you want to replace these with solid wood planking, then this should be of ¾in (19mm) lumber, because solid wood is more likely to warp than firwood. All the firwood panels can be cut from three large sheets as shown in Figs. 1 to 3.

Where screw heads are visible, then brass screws with countersunk washers are used; but where they are not, ordinary flathead steel screws are just as good.

Cutting the panels

All the ½in (13mm) thick firwood panels can be cut from three large sheets of firwood, and the cutting plan for this is as described. This will avoid the cost of the panels being cut to specific sizes. But this only applies if you have done some carpentry and have access to a bench power saw. Cutting long straight lines in wood is no job for the inexperienced carpenter. So if you have never done this sort of work, order the firwood already cut into eight individual panels according to the dimensions shown in Figs. 1 to 3.

First prepare the two base side panels. Each panel has two recesses cut in one end as shown in Fig. 4. The recesses are 5¼in (133mm) wide and 3½in (89mm) deep. Mark out the outlines of the recesses, using a try square to ensure that all lines are straight, then cut out the recesses with a fine-toothed saw. When you have finished, you will have two identical panels like the one shown in Fig. 8.

Next cut the recesses in the foot board. These recesses are identical to those cut in the side panels, but in this case they are cut in each end of the foot board. Stand one of the side panels on edge, in the position shown in Fig. 8. Do the same with the foot board, as in Fig. 9, and then mark the positions of the 'tenons' on the footboard and jig-saw the two together so that the protruding parts of one panel slide into the recesses in the opposite panel. Now add the remaining side panel to the opposite end of the foot board in the same way.

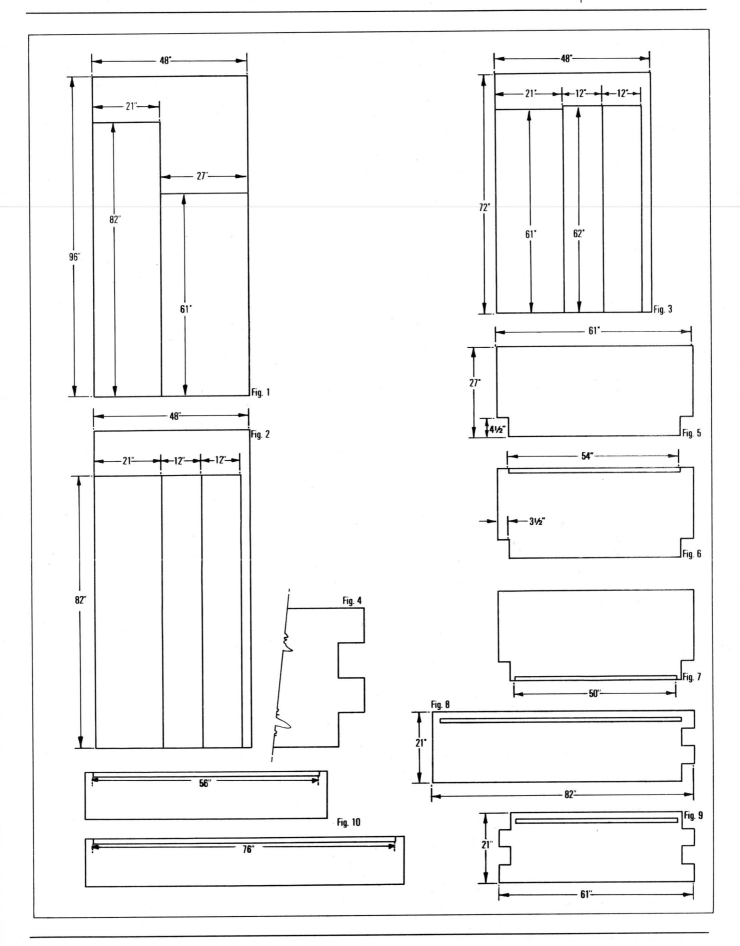

Fig. 1
Fig. 2
Fig. 3
Fig. 4
Fig. 5
Fig. 6
Fig. 7
Fig. 8
Fig. 9
Fig. 10

Fix the two 73in (1853mm) battens to the side frames, 2½in (64mm) down from the edge of each inside face, and the 54in (1371mm) batten in the same position on the foot board. Figs. 8 and 9 show the battens in position.

Mark and cut the two recesses in each bottom corner of the headboard. Each recess is 3½in (89mm) wide and 4½in (114 mm) deep, as shown in Fig. 5. Glue and screw a 54in (1371mm) batten along the top of one side of the headboard, and a 50in (1270mm) length along the opposite side, at the bottom. Fig. 6 and 7 show these in place.

Fix one length of 76in (1930mm) battening along one side of each of the top side panels. The batten should be flush with one edge, and centrally placed so that there is a space of 3in (76mm) between each end of the batten and the ends of the panel the same as with the back of the headrest. Repeat this with the top end panels and the 56in (1422mm) battens.

Assembling the main frame

You are now ready for a dry run, to check that everything fits, and to mark and fit the slats. Do not apply glue to any of the joints at this stage and do not drive the screws in too tightly. This is because each panel will have to be removed, covered with felt, and replaced before the slats are finally screwed in place and the bed is ready.

Assemble the two base side panels and footboard so that the three panels are free-standing.

Where the panels interlock, the ends will protrude, forming a V at each corner. Stand an upright in one of these, so that the panel ends overlap two sides of the upright, then screw them together through the panelling into the upright. Repeat this at the other end. The uprights for the foot end of the four-poster are now in position.

Now fit the uprights at the head end. Stand an upright at the outside end of a side panel, in exactly the same position as the upright at the opposite end. Screw this in position through the inside of the panel. Repeat this procedure with the remaining upright on the opposite panel. Stand the headboard in position on the edges of the side panels, backing onto the uprights, and screw it to the uprights through the front of the headboard.

The base frame is now assembled, and you can check the various panels for fit. You can also check the top frame. It is not necessary to assemble this; it should be sufficient to hold each panel up (you will need someone to help you) to check that it fits.

Mark, cut and fit the slats. Do this by direct marking, and as you cut and fit each one, number it on the slat and the butting batten, so that each slat will be fixed in the right place when you assemble permanently. Do not screw the slats in place—it is only necessary to drop each one into postion to see that it fits. Carefully dismantle the unit. The panels can now be covered with felt; and the bed can be assembled permanently and, if required, rods can now be fixed around the inside of the top frame to hold curtains to enclose the bed.

Cutting list

Solid wood	standard	metric
4 uprights	3×3×84	76×76×2134
11 slats	1×6×54	25×152×1371
2 slat supports	2×2×73	51×51×1853
2 edge battens	2×2×54	51×51×1371
1 edge batten	2×2×50	51×51×1270
2 edge battens	2×2×76	51×51×1930
2 edge battens	2×2×56	51×51×1422
Plywood		
2 basic side panels	½ × 21 × 82	13 × 533 × 2083
1 foot panel	½ × 21 × 61	13 × 533 × 1549
1 headboard panel	½ × 27 × 61	13 × 686 × 1549
2 top side panels	½ × 12 × 82	13 × 305 × 2083
2 top end panels	½ × 12 × 62	13 × 305 × 1575

You will also require:

Brass screws, with cups for the visible screw heads. Ordinary countersunk head steel screws can be used for screws that are not visible, such as the securing screws for the slats. A good woodworking adhesive will ensure firmer joints and a better finish.

Linen chest

The linen chest is made from high quality polished pine with the exception of the base, which is made from plywood. The sides, front and back are dovetailed together at the corners. The lid rests on two pieces of pine between the sides. The base rests on similar pieces glued to the insides of the front and back at the bottom of the chest.

Preparation of the wood

The sides of the chest are 18in x 18in (457mm x 457mm) and the front and back are 14½in x 38in (368mm x 965mm). The four panels of the chest are made from the same plank because it is almost impossible to achieve a precise color match if different planks are used, even if they are from the same tree. Using wood from the same plank will also enable you to achieve the decorative effects.

First, cut the 9ft 4in (2.8m) length of lumber down its thickness. Use a circular saw, allowing the rip fence to guide the accuracy of the cut. It is best to use a combination blade but since the lumber is to be planed afterward, a rip blade can be substituted. You will have to turn the lumber over after the first cut, and cut it again from the other edge. Alternatively, ask your lumber dealer to saw the planks. By cutting the lumber in this way you get two matching planks. When the cut is completed, 'open' the two pieces as you would open a book (Fig. 2). The grain in one plank will be perfectly matched, or mirrored, with that in the other plank.

Plane the planks to the required thickness of ¾in (19mm) and the bottom edges, with the matching grain, exactly square. Use a jack plane or router to do this. Check that the faces and edges are level with a straightedge, using a try square to check that they are exactly at right angles to each other.

Next, butt the planed edges with the matching grain together. Mark them as shown in Fig. 3. This will enable you to match the grain easily later on.

With boards still together square a line down the face close to one end. From this line measure the length of the components in the following sequence: front 38in (965mm), side 18in (457mm), back 38in (965mm), side 18in (457mm). Add ¼in (6mm) to each of these measurements to allow for waste. Marking the panels in this way will enable you to achieve another decorative effect, as the grain on the finished piece will appear to flow around the sides of the chest.

Gluing the planks

To form the components of the chest glue the planks together along the planed edges. You will need a number of bar clamps—at least one for every 18in (457 mm) of the length of the planks—and a PVA adhesive. If you do not have enough bar clamps or room to work with two planks as long as this, cut them across their width down the line that marks off one set of side and end panels from the other.

To enable you to join the planks quickly once the adhesive has been applied, pre-set the clamps to the correct length by clamping the boards together dry with blocks of waste wood between the jaws of the clamps and the edges of the planks. This will protect the edges of the planks from damage and will spread the clamping pressure more evenly. If you have cut the boards across their width, three clamps is the minimum number to use, although four clamps would be ideal.

Release the clamps and spread adhesive evenly on the edges of the planks that are to butt. Clamp them together again with the bars of the clamps alternatively on the top surface of the panels and on the underface of the panels.

It is essential that the planks fit together perfectly and the surface is completely level. If one plank is slightly higher than the other slacken the clamps a little. Place a piece of scrap wood on the raised plank and bang it with a hammer. This will force the raised plank downward, until its surface is flush with the other plank.

Tighten the clamps and wipe every trace of excess adhesive off the face of the planks with a soft damp cloth. Run your finger along the joint to check that there are no irregularities in it. Allow adequate time for the adhesive to set—12 hours is usually sufficient, but follow the manufacturer's recommended times. It is better to be overcautious at this stage than to spoil the joint and have to go back to begin to do the work all over again once it's been ruined.

When the adhesive has set, cut the joined planks along the marked lines. Plane the two shorter lengths (the sides) to a width of 18in (457mm). Then take the two longest pieces (the front and back) and mark off their final width of 14½in (368mm). To do this, measure half this distance from the joint and on both sides of it. (This will keep the joint, and grain, running continuously, around the finished box.) Draw lines through these points parallel with the join. Then cut the wood to the waste side of these lines and plane the planks square and to width.

Making the dovetails

The side panels are joined to the front and back panels with dovetailed joints. But first, you must cut and plane the boards to their final length. From one end of each board mark off approximately $\frac{1}{8}$in (3mm)

Fig 1. Construction details of the linen chest. Note the position of the battens which prevent the lid from warping.

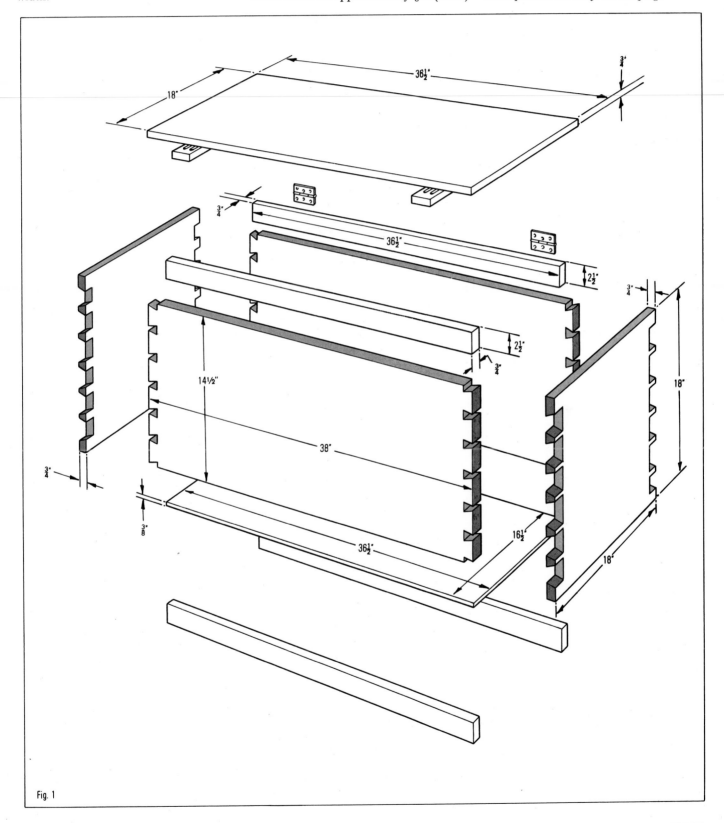

Fig. 1

Linen chest

A large plane is used to smooth the planks.

Method when planing end grain.

Bar clamping the planks together.

Make sure that the planks fit together perfectly and the surface is level.

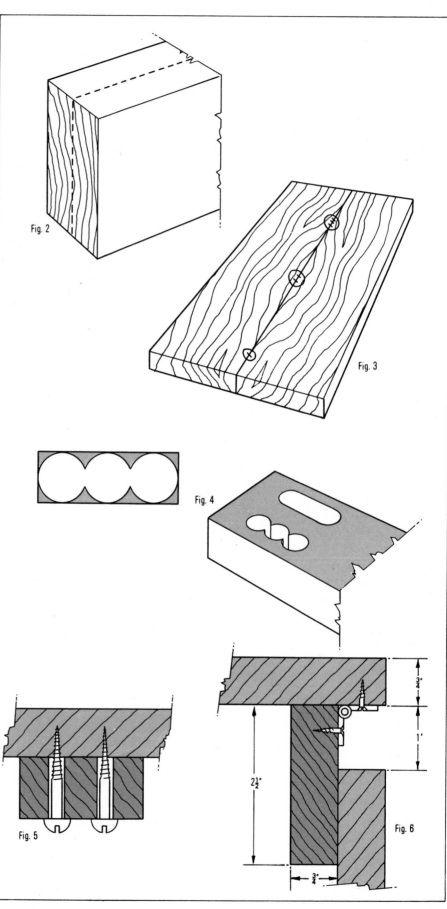

Fig. 2

Fig. 3

Fig. 4

Fig. 5

Fig. 6

mark a line to indicate the depth of the dovetails. Set a marking gauge to the width of the wood. Score the line inwards from both ends of all four boards. Mark this line very lightly, as it will have to be planed out later.

Then measure and mark a point $1\frac{3}{4}$in (45mm) down from the top of both side pieces. Do this at each end. This will indicate the point at which the top of the front and back pieces join the sides. Mark out the dovetail between this point and the width of the front and back panels. Cut the tail (wider pieces) of the dovetails first. Then direct mark and cut the pins. Ensure that the tops of the front and back pieces are level with the point marked on the sides. Trial assemble and then glue the sides together.

Refer to the Techniques section for the two methods of cutting the dovetail. By using a dovetail attachment and your electric drill you will achieve much greater accuracy, although you may feel that this is not worth the expense just for one project. In which case, try the jig saw method, practising on a scrap piece of timber first.

Making the lid

The lid for the chest is $36\frac{1}{2}$in (927mm) long and 18in (457mm) wide. Two planks have to be butted together to form a surface this wide. Cut the timber for the lid so that it splits down the middle. Plane the edges and butt and glue them to form a plank in the same way as for the side components.

A board of this size which is fixed along one edge is unlikely to stay perfectly flat and level. The underside is, therefore, strengthened with battens. Cut two 2in × 1in (51mm × 25mm) battens to a length of 12in (305mm). The battens are not screwed in the normal way. Instead, the holes through the battens which accommodate the shanks of the screws must be elongated to allow for any movement in the timber caused by variations in temperature and humidity.

Mark the positions on the battens through which the screws will pass, four pairs of screws for each batten is sufficient. Drill holes at these points. Drill another hole of the same size on either side of every original hole (Figs. 4 and 5). Chisel out the timber around each pair of holes to form a rectangular opening.

Use round head $1\frac{1}{2}$in (38mm) No.8 brass screws to fix the battens to the lid. Before using these screws pre-cut the thread in the lid with steel screws. This ensures that when you screw in the brass screws they will not snap in the hole and the head will not be damaged.

Finger hold and scuff board

To make the finger hold and the scuff board and their balancing pieces at the back you will need four pieces of $2\frac{1}{2}$in × $\frac{3}{4}$in (64mm × 19mm) softwood or struts. Cut and square these to a length of $36\frac{1}{2}$in (927mm). Glue the two struts at the top to the insides of the front and back panels so that 1in (25mm) of their width is visible from the exterior of the chest. Position the bottom struts so that 2in (51mm) is visible. The base of the chest can now be cut to rest on the bottom struts and is glued into place.

The lid of the chest rests on the top struts and between the side panels, flush to their tops. Plane the lid to the required size, but ensure that the fit is not too tight. If you can slide a thin coin between the end of the lid and the sides the fit is ideal.

Fix the lid to the back strut with three 2in (51mm) brass butt hinges and some $\frac{1}{2}$in (13mm) brass screws. The plates of the hinges when fixed are at right angles to each other (see Fig. 6). The exact position of the hinges is not really important provided one is in the middle and the other two are placed near the ends.

Finishing the chest

Clean up the surfaces with a sharp smoothing plane and sand them perfectly smooth with your orbital sander, or by hand, using fine paper. Use a clear polyurethane or cellulose varnish so as not to obscure the grain. Rub down the finish after each coat with steel wool until you have achieved a completely smooth finish.

Fig 2. If you have the facilities for cutting a 9in (229mm) plank through its thickness or can get the job done commercially, the components for the chest may be made from a plank $1\frac{3}{4}$in (44mm) thick.

Fig 3. The two resulting planks should be opened like a book and the grain aligned for glueing as shown. The effect created will be of a mirror image which continues round the chest.

Fig 4. The under lid battens require slots to allow for expansion of the wood. These are cut as shown, using a drill and subsequently a chisel to remove the waste.

Fig 5. Brass round head screws are used to fix the battens to the lid allowing the wood to slide if there is any expansion.

Fig 6. The method of hinging the lid so that the back provides its own support.

Cutting list

Solid wood

	standard	metric
2 planks for sides, front and back	$112 \times 9 \times \frac{3}{4}$	$2845 \times 229 \times 19$
2 lids	$36\frac{1}{2} \times 9 \times \frac{3}{4}$	$927 \times 229 \times 19$
2 battens	$12 \times 2 \times 1$	$305 \times 51 \times 25$
4 scuff board, finger holds	$36\frac{1}{2} \times 2\frac{1}{2} \times \frac{3}{4}$	$927 \times 64 \times 19$

Plywood

1 base	$36\frac{1}{2} \times 18 \times \frac{3}{8}$	$927 \times 457 \times 10$

You will also require:

3 2in (51mm) solid brass butt hinges. $1\frac{1}{2}$in (38mm) roundhead brass screws. $\frac{1}{2}$in (13mm) No.8 countersunk brass screws. PVA adhesive.

Garden bench

A garden bench can be one of the most attractive features of any garden. Unlike a lot of garden furniture, this design is not only attractive and comfortable, it is also sturdy enough to be used as a permanent feature all year round and to give a lifetime's use.

The garden bench has been designed specifically for do-it-yourself construction. No special tools other than a protractor, are required and all the joints are simple. At the same time the bench is very sturdy and, provided the lumber has been properly treated with preservative, is suitable for outdoor use throughout the year.

One major attraction of the design is that it can easily be modified to make a bench of a different size, or even an individual garden chair. The construction steps would remain exactly the same, the only difference being that the main members for a chair should be reduced in section from 2in x 4in (51mm x 102mm) to 1½in x 3in (38mm x 76mm). By adapting the design you can build a complete set of matching garden furniture.

Choosing a suitable wood

There is a wide range of woods suitable for use. Among the softwoods, pine is the most readily available and is easy to work. It requires treatment with an exterior wood preservative, even if you intend to paint the finished structure. Of the hardwoods, elm, oak, teak and makore are the most suitable, but they all have different properties and the home carpenter should make his choice after a consideration of their individual characteristics. Oak is strong and highly weather resistant, but its extreme hardness makes it difficult to work. Elm is suitable for most outdoor uses—its only drawback is that it tends to split more easily than the other woods listed. Perhaps teak combines all the most desirable properties. It is a very attractive wood, which is relatively easy to work and, unlike the other woods, is so oily that it does not need any special treatment with a preservative. Unfortunately it is extremely costly, although this high initial cost is offset by its long life.

Treating the lumber

Since the wood will be used outdoors, it is a good idea to treat it with a preservative, at least those portions in direct contact with the ground. Some woods, such as redwood may be left alone and will stand up to the weather quite nicely. If you wish, you can use pressure-treated lumber, which is pre-treated with preservative.

The end sections

Each end section comprises an armrest a front and back leg and a rear, vertical support These members are assembled by gluing and jointing; final securing being by galvanized lag screws countersunk below the surface When finished, these two sections should match exactly, so great care must be taken in marking and cutting out

Begin by marking and cutting out the pieces for the two end sections to the dimensions given in the cutting list. All these pieces with the exception of the armrests, are ½in (13mm) overlong at this stage, to enable the angled ends (shown in Figs. 3 and 4) to be cut out accurately.

Now take one of the front legs and mark a point 22in (559mm) along one long edge. With the aid of a protractor, set a sliding bevel gauge at an angle of 83°, lay the gauge against the marked side and draw in a line from the measured mark to the opposite edge. Cut along this line with a crosscut saw to give the completed leg. Repeat this procedure on the other front leg member and, when it has been cut out, match the 'paired' legs and, if necessary, trim them so that they are of an identical size.

The rear legs are angled at both ends. To cut them accurately, first mark off the overall length plus a little extra with a try square. Lay the sliding bevel, pre-set to 70°, to coincide with one of the marks and draw in the angled line. Reset the sliding bevel to 75°, and mark in this angle from the diagonally opposite marked point. Now cut along the angled lines to give the two rear legs.

Next, take the two rear, vertical support members, and mark a line ½in (13mm) from one end of each. With the sliding bevel set at 102°, draw in the angled ends as detailed previously and cut the members to shape. Cut out the top face with a jigsaw or band saw, according to the pattern shown in Fig. 4. Do not cut out the halved joints at the bases of these members yet, instead, cut out the other component pieces of the end sections.

Take the end rails and, using a try square, mark a line ½in (13mm) from one end of each. Set your sliding bevel to 70° and, referring to Fig. 4 as a guide, mark in the angled ends. Cut these ends out with a crosscut saw. At the opposite ends mark out and cut the bevels shown in Fig. 3. The upper long edges of the end rails are hollowed out with a coping saw or jigsaw to the shape shown in Fig. 3.

The last parts of the end sections to be cut out are the arm rests. Round off the last 2in (51mm) of the front ends of these pieces and, with a spokeshave, make 'nosings' on these ends. At the other ends, cut the lapped joints which house the vertical supports.

Cutting the end section joints

The rear support and the end rail are joined together by means of a halving joint. Although this is an easy joint to make, in this case the process is complicated by the fact that the two members are set at an obtuse angle to each other. Refer to Figs. 3 and 4 for the exact measurements and when you are satisfied that you have marked out correctly, cut out the joint.

Both the front and top horizontal rails are let into the end rail and rear support respectively. These dadoes are ¾in (19mm) deep and the width of the 2in x 4in (51mm x 102mm) lumber. Their exact locations are given in Figs 1 and 4.

When all the joints have been cut, trial assemble all the pieces and, if necessary trim for a good fit.

Assembling the end sections

Initially these pieces are joined by gluing and nailing, final fixing being by lag screws. Begin by gluing the front and rear legs to the armrests according to the plan shown in Fig. 4. Then glue the end rails inside the legs so that their angled ends lie flush with the long outside edges of the rear legs, and their lower long edges meet the outer edges of the front legs at a point 10¾in (273mm) from their bases. To complete the assembly, glue the rear supports into the recesses on the armrests and end rails so that their lower ends are flush with the bottom edges of the end rails. When the glue has dried, use 10d galvanized nails, driven in from the inside faces, to secure the structure. Do not fix with screws just yet.

The seat and back

Three rails connect the end sections. One of these is glued and screwed into the housings on the front of the end rails, another is butted to the rear of the

Garden bench

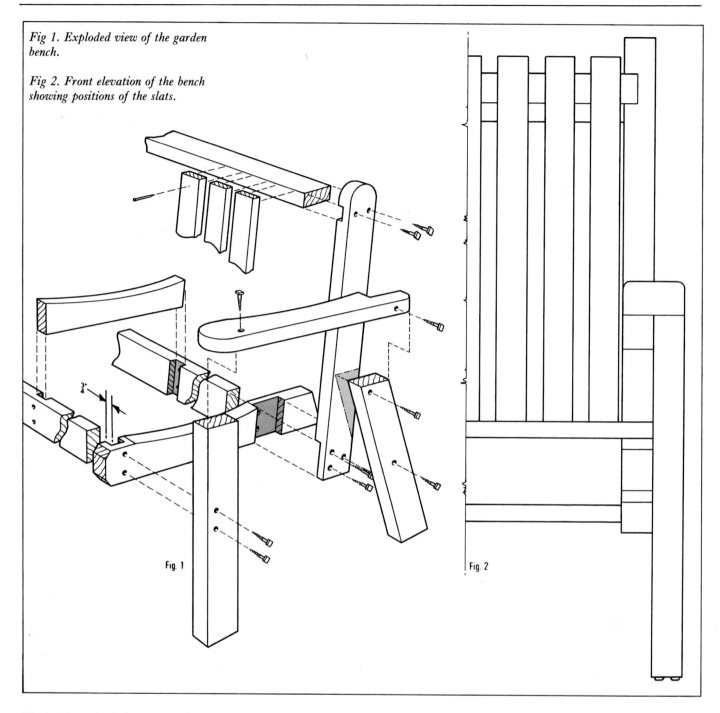

Fig 1. Exploded view of the garden
bench.

Fig 2. Front elevation of the bench
showing positions of the slats.

Fig. 1

Fig. 2

$\frac{3}{4}$"

Fig 3. The ends of the seat are shaped
as shown. The inside curve at the top
of the component may be cut with either
a spokeshave or a rasp.

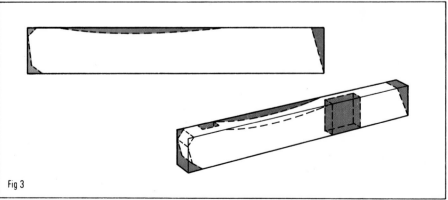

Fig 3

Cutting list

Solid wood	standard	metric
Main frame lumber	2 x 4 x 408	51 x 102 x 10,363
Back slat lumber	$\frac{3}{4}$ x 4 x 312	19 x 102 x 7.924
Seat slat lumber	1 x $1\frac{3}{4}$ x 420	25 x 45 x 10,668

Cut to give:

2 front legs	2 x 4 x $22\frac{1}{4}$	51 x 102 x 565
2 rear legs	2 x 4 x 22	51 x 102 x 559
2 arm rests	2 x 4 x 23	51 x 102 x 584
2 end rails	2 x 4 x 26	51 x 102 x 660
2 vertical supports	2 x 4 x $31\frac{1}{2}$	51 x 102 x 800
1 front rail	2 x 4 x $57\frac{1}{2}$	51 x 102 x 1,460
1 top rail	2 x 4 x $57\frac{1}{2}$	51 x 102 x 1,460
1 back rail	2 x 4 x 56	51 x 102 x 1,422
1 center rail	2 x 4 x 16	51 x 102 x 406
11 back slats	$\frac{3}{4}$ x 4 x 28	19 x 102 x 711
7 seat slats	1 x $1\frac{3}{4}$ x 60	25 x 45 x 1,524

You will also require:
18 6in (152mm) galvanized lag screws. 6 3in (76mm) lag screws. 2 $1\frac{1}{2}$in (38mm) galvanized lag screws. 1 lb (0.45kg) 2in (51mm) galvanized nails. $\frac{1}{2}$lb (0.2kg.) 3in (76mm) galvanized nails. Waterproof woodworking glue. Waterproof stopper. Wood preservative. Sandpaper.

end rails in line with the front of the back supports and the third slots into the dadoes cut in the top of the vertical supports. Before fixing them in postion, cut the dadoes for the center rail at the midpoints of the two seat rails. These slots are 1in (25mm) deep and the width of the center rail.

At this stage, cut the center rail to shape using Fig. 1 as a guide. Use a coping saw or jigsaw to hollow out the upper edge of the center rail and check that this hollow corresponds exactly to those cut on the end rails. Now glue and nail the front rail into position.

When the glue has set, fit the back rail. This piece is fixed between the end rails at the same angle as the vertical supports; the center rail follows. The exact location of the back rail is shown in Fig. 4. Finally, glue the top rail into the housings on the vertical supports.

Fixing with lag screws
Figs.1 and 4, show the location of the lag screws. Use 6in (152mm) screws to secure the end section members to the main rails, and 3in (76mm) screws at all the other joints. Pre-drill for the screw threads, taking care to avoid nails, and allowing an extra $\frac{1}{4}$in (6mm) for countersinking the screw heads. Insert the screws and fill the exposed cavities with wood plugs which can be cut either from waste wood with a plug cutter, or, from dowel of the correct diameter.

Adding the slats
This is the easiest part of the construction. Begin by nailing the vertical rear slats to the rear and top rails, after marking the center location of each. To help you space the slats evenly, allow about $\frac{7}{10}$in. (18mm) between each one.

Now nail the horizontal seat slats in position, allowing about $\frac{1}{2}$in (13mm) between them. With a spokeshave, round off the edges of the seat slats and make a 'nosing' on the leading edge of the front seat slat. Finally, smooth the whole structure with sandpaper and apply a final dressing of preservative against weather conditions.

Your garden bench is now complete: the work involved will be amply rewarded by the pleasure the bench will give throughout the year, both as an object of beauty and a practical addition to your garden.

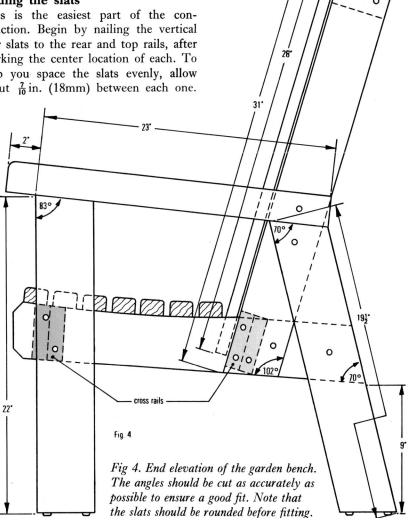

Fig. 4

Fig 4. End elevation of the garden bench. The angles should be cut as accurately as possible to ensure a good fit. Note that the slats should be rounded before fitting.

Garden seat/planter

This wood seat planter makes an attractive addition to any garden, either as a major feature in a small area, or as part of a large landscaped garden. The center of the unit can be used to display flowers or shrubs, or you can build the seat around a tree trunk so that the branches will provide shade.

General constructions details
The unit is a hexagonal or six-sided garden seat, the center of which has a well that is filled with soil and planted out with a display of your choice.

The main framework consists of 1in x 2in (25mm x 51mm) softwood members, clad with ¾in x 4in (19mm x 102mm) cedar wood tongue-and-grooved (T&G) treated boarding; you will require sixty lengths of 1ft 8in (508mm) boarding for the base, and eighty lengths of 9in (229mm) for the top. All other wood members, such as the seat boards, soil support and base corner plates, are of either 1in or ½in (25mm or 13mm) marine (this is most important) plywood.

All these parts must be treated with a horticultural grade of wood preservatives such as Cuprinol. In addition the internal top lining of the T&G boarding, and parts O, H and J must be coated with a bitumen-based preservative such as Creosote to prevent damp soil from rotting the wood.

The outside supports consist of six outer base frames D joined with halving joints (Fig. 3). These form the outer hexagon. Six inner base frames E provide the internal support, and these radiate from the middle and are joined to the outer frames as shown in Fig. 14. The inner frames are constructed with bridle joints to provide greater strength vertically.

The inner hexagon, which houses the soil, is formed by six rectangular frames, clad both sides, and with an internal base H of $\frac{1}{2}$in (13mm) marine plywood (Figs. 1 and 9).

If you prefer to build the seat around a tree, then you will have to omit the center base plate, and replace it with six blocks of $\frac{1}{2}$in (13mm) thick marine ply,

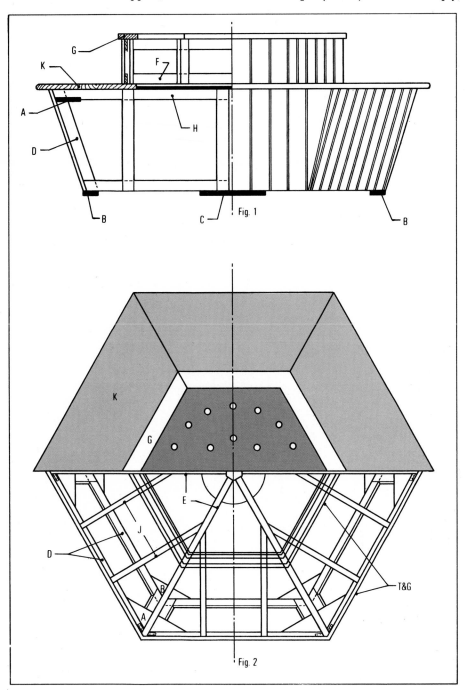

Fig 1. Side elevation of the garden seat and planter.

Fig 2. Plan view of the seat. If the unit is to be constructed around a tree the center section will require a certain amount of modification.

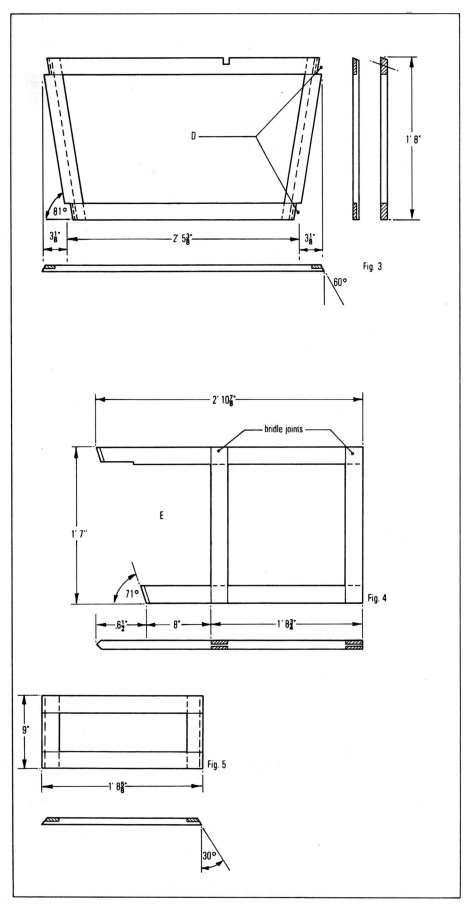

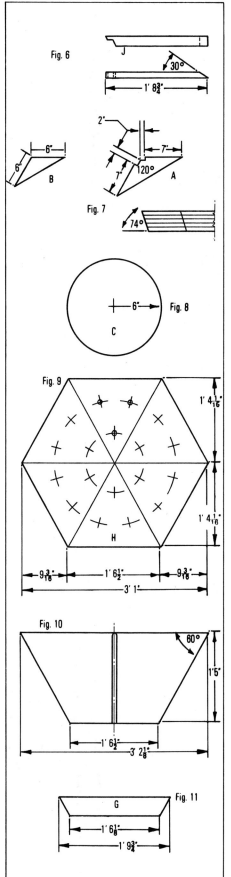

one under each inner base frame, to keep the structure off the ground and to keep it level with the base corner plates. The soil container will obviously have to be left left out of the design. These alterations will not weaken the unit, because the missing pieces are needed only as a support for the soil.

All joints must be screwed—use brass or galvanized screws to prevent unsightly rust stains—and glued with a waterproof woodworking adhesive such as Resorcinol. Remember, though, that not all combinations of wood preservative/wood adhesive will allow a secure join. Some adhesives, for instance, will adhere well to wood that has been treated with Cuprinol clear, or light oak, but not to wood that has been treated with Cuprinol red cedar. Your local supplier will provide you with details from the data sheets that the manufacturers send out.

The ground underneath the unit must be level. Otherwise it will be set as an unsightly angle and you will be sitting on a sloping seat. You will usually be able to level the area with a shovel. However, if the ground slopes too much you might have to raise the structure on a concrete foundation.

Outer base frames

First make up the six outer base frames as shown in Fig. 3D. Each frame consists of four members, with half-lap joints at each corner. The edges of the two sides or vertical members are beveled or mitered so that the frames will butt together and form the hexagonal shape shown in Fig. 2.

The joint has to be angled, because the corners are not square. This is done using the same technique, except that the angles are marked out using a bevel gauge.

The next step is to miter or bevel the edges. With the bevel gauge, mark out the angle on the end-grain of each end, then draw with a marking gauge and pencil a line along the inside of the lumber between each marked point. Cut the wood down to this line, preferably using a table saw. If one is not available, the angle can be cut with a rip saw. Place the lumber vertically in a wood vise. To prevent the wood from vibrating while sawing, start with only about a 10in (254mm) length of wood above the vise and gradually raise the member as the cut is made. The last few inches can be cut by reversing the mem-

ber in the vise and continuing from the opposite end. If you do not have a power saw or a rip saw, you could plane the wood away—but your arms will ache when you have finished.

Assemble the frame and check that the angles are correct. To do this, measure the side members and see that they are identical in length, then measure the diagonals. Each measurement must be identical. Glue and screw the joints and leave for the adhesive to set.

When the glue has set, cut the notches to accept the ends of the inner base frames. (The recess on the top that will take the outside ends of the seat support members J [Fig. 2] can be cut later after marking out by direct marking.) Make up the other five frames in the same way.

Inner base frames

The construction outline of these is shown in Fig. 4E. Each frame consists of a top and bottom member, and two vertical members, one centrally placed and one at the side, creating, in effect, one open and one closed side. The open end of the frame is housed into notches cut into the corners of the outer base frames.

The frame is secured with bridle joints. These are the best joints to use when vertical members are under compression—as these will be when supporting the amount of soil required.

The bridle joints, which are like an open mortise-and-tenon joint, should be marked with a mortise gauge, which is similar to an ordinary marking gauge except that it has two scribing points or spurs, one of which is adjustable to give varying widths of mortise or tenon. To set the gauge, first loosen the stock setting screw (Fig. 3) and slide the stock back. Next adjust the width—in this case $\frac{5}{16}$ (8mm)—and set the gauge so that the distance between the points locates in the center of the member. Check that it does, by measuring the distance from each scribed line to the edge of the member. Each of these measurements must be identical for accuracy.

After marking out, cut the joints. Cutting techniques are similar to those used for a mortise-and-tenon joint. Trial assemble the frame and check the diagonals for squareness, then glue and screw. Do not worry too much about the lengths of the top and bottom members at the open end. They can be finished off by direct marking at a later stage.

Fig 3. The outer frames of the garden seat are built to this pattern. A total of six is required. Note the angle at the top edge of the units.

Fig 4. Six frames to pattern E are required. Note the shape of the ends of the frames. These butt to the ends of frames D.

Fig 5. The inner frames of the planter itself. Again, six are required to pattern F.

Fig 6. Pieces J form the braces and seat supports and are cut to pattern J.

Fig 7. Triangles A and B are the corner braces. These are cut to the pattern shown. Piece B fits at the angle which joins pieces F. Piece A fits at the joint between pieces E and F underneath in the cut-out provided in piece E. Note the angles cut to accept the uprights of D.

Fig 8. Piece C is the center base plate. It is not absolutely necessary that this piece be circular.

Fig 9. The base of the planter. It is this component which carries the soil and drainage holes must be drilled so that excess water can drain away.

Fig 10. The pattern for the seat panels.

Fig 11. Pieces G are the top combing of the planter.

Garden seat/planter

Opposite page left to right.

Top row: The method of using the marking gauge for marking the bridle joints.

The joint is cut with a tenon saw and the waste chopped out with a chisel.

The other component of the bridle joint. Again the first cuts are made with a tenon saw and the waste is removed with a chisel.

Middle row: Completed frame E. The bridle joint is used to retain the strength of the vertical members.

The method of marking the bevels at the ends of pieces D.

Once the angle has been marked on the end grain, the gauge is set to the measurement and a line is scribed down the length of the upright.

Bottom row: The waste is removed, either with a bench power saw or with a jack plane.

Close-up of the finished basic frame before the corner braces are fitted.

Plan view of the completed frame.

Top frames

These are simple rectangular frames with a halving joint at each corner. Full construction details are shown in Fig. 5F. These frames form the walls of the soil container and are boarded over on the outside to provide the seat back. They can be built at a much later stage.

Other components

The center base plate is just a disc of $\frac{1}{2}$in (13mm) marine plywood as shown in Fig. 8C. It is not essential for this piece to be circular, and if you want to save time you can just as effectively use a 12in (305mm) square of plywood.

You may, however, prefer the neatness of the circle. In this case lay a piece of 12in x 12in (305mm x 305mm) ply on a table. Lightly tap a nail about one third of the way into the center of the board, and tie a 6in (152mm) long piece of string to the nail. Tie the other end to a pencil, stretch the string out and draw a circle on the board. Then cut out the circle with a coping or jigsaw.

The soil support is the hexagonal base for the soil and is shown in Fig. 9H. It is cut from a 2ft 8$\frac{1}{8}$in x 3ft 1in (816mm x 940mm) sheet of $\frac{1}{2}$in (13mm) plywood. Lay the ply on a level surface and mark a line lengthwise down the middle. Along each *long* side *only*, mark off a point 9$\frac{3}{16}$in (233mm) away from each corner. This will give you two marks along each long edge. Mark a line from each of these marks to the nearest end of the the middle line, and you have your hexagon ready for cutting out, in four cuts.

Corner plates are triangles of wood that provide bracing for the outer frames. They are easily cut from sheets of $\frac{1}{2}$in (13mm) plywood by following the diagrams in Figs. 7A and B.

Coping pieces are run around the rim of the top frame to protect the end-grain of the boarding and conceal the boarding/framework joint. Each one is cut from a 1ft 9$\frac{3}{4}$in (552mm) length of $\frac{1}{2}$in x 3in (13mm x 76mm) ply as shown in Fig. 11G. Seat supports—the cutting details for these pieces are shown in Fig. 6J—are housed at one end into the top of the outer frames, and butted at the opposite end, at an angle, to one side of the top member of an inner frame and screwed in place.

The seats are cut from 1in (25mm) ply as shown in Fig. 10K. The top surface should be lightly sanded down, and the front edge rounded slightly with a plane to provide comfortable seating.

Assembly

There are several ways in which the unit can be constructed, but the following method is one of the easiest. You begin by building part of the unit upside down.

Place the soil support on level ground. Mark out the frame positions on it, place two adjacent inner frames in position and skew-nail or toenail them in position (leave the heads protruding slightly in case the positions need adjusting). This will allow one of the outer frames to be fitted in position. Repeat this procedure until the hexagon is complete, gluing and screwing each piece into place when you are sure of the fit. Each outer frame is joined to its adjacent outer frame by screws through the inside of the side members. While the unit is still upside down, screw the center base plate and all the base corner plates in position.

Carefully turn the structure right way up and, using direct marking, mark out, cut and fit the seat support members. Trial assemble the top frames around the soil support and when you are sure that the fit is correct, screw the top frames to the tops of the inner frames, and to one another as for the outer frames.

Trial assemble the seats. When you have made sure they fit (by planing a few edges, if necessary), secure them by screwing from underneath through the inner base frames or seat supports.

Mark out, cut and fit the boarding for the bottom section. Toenail each T&G board through the tongue at each end so that the nail heads do not show. Repeat this for the top section.

Trial assemble the coping pieces around the top lip of the well and, when the fit is accurate, toenail them in position using finishing nails, punching the heads down.

Filling the planter

The well of the seat planter can now be filled with soil. If you can afford to do so, use one of the peat-based potting soils. These are only a fraction of the weight of soil and, apart from easing physical strain, will lessen watering problems as peat retains more water than soil.

If you intend growing shrubs in the well, make sure that you do not plant ones that tend to deep-root. The depth of the growing medium is only about 9in (229mm)—unsuitable for some shrubs.

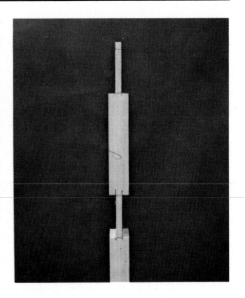

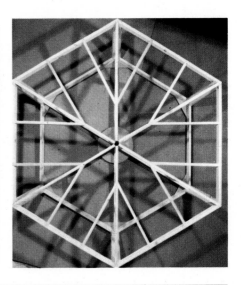

Garden gates

Wooden garden gates, however well made and painted, inevitably decay and need replacing after some years. Lumber dealers generally stock replacements in a range of sizes. Unfortunately, gate posts, particularly those of old houses, do not always conform to the standard sizes. If you want a new gate, you will either have to have one built, which can be expensive—or make one yourself.

If you choose to make your own gate, you will have complete control over the design instead of having to make do with what you can buy. A few variations on a simple, basic design are given here, but all use the same woodworking techniques, and so will almost any version that you design yourself.

A garden gate has to stand up to considerable wear and tear, including children swinging on it. So it must be sturdily made, using the same standards in terms of material and careful workmanship as you would use for indoor furniture.

Materials

The cheapest way of making a gate is to use ordinary softwood. Provided it is properly primed and painted, it should last for years. But the more you are prepared to spend on lumber, the longer-lasting the job will be. Oak, at three or four times the price of softwood, and teak, which is even more expensive, are particularly durable.

The gate design given here uses tongue-and-groove boarding, which is often not obtainable in hardwood. Fortunately, its use is not essential. You can use plain boards, or build a gate of a design not using boarding at all. Refer to the cutting list for a list of all the materials used for the gate.

The quantities given in the cutting list are for a gate 4ft (1219mm) wide and 3ft (914mm) high, and suitable for an opening 4ft 1in (1225mm) between the gateposts. Use planed lumber which will be slightly narrower than the dimensions stated.

It is also important to choose the right nails and hinges. Nails used for outdoor work should be galvanized to resist rust, and screws should be of the black japanned type. This is particularly important if you are making a hardwood gate, because the natural acids in the wood attack bare steel. Use oval head screws 1½in (38mm) long.

The hinges for the gate can be of several types (Fig. 3). Ordinary tee hinges up to 18in (457mm) are quite strong enough for a medium-sized gate. A larger gate—more than about 5ft (1.5m) wide—will need the stronger cast-iron hinges. These are available with normal fittings for wooden posts or special flat plates for bedding in the mortar joints of brick pillars. This gate would need 2ft (610mm) bands and hooks or 2ft 6in (762mm) if it is made in heavy hardwood such as oak.

If you want your gate to fold back flat against the wall, use Parliament hinges which have offset pivots.

Measurement and planning

The first step is to measure the site and note the condition of the existing gateposts. They will probably need replacing at the same time as an old gate. If you are installing a new gate between existing posts, the total width of the gate must be about ½in (13mm) narrower than the space between the posts, and perhaps more for a large gate. This gap allows for

the fact that at a slight angle (say, as you begin to open it) the gate needs extra room if it is not to jam, as well as allowing for the inevitable winter swelling. (Doors, but not gates, are bevelled to allow for this.) If you are using Parliament hinges, the gap should be proportionally larger.

Once you know the height and width of the gate, draw up a plan of the design you want. This need not be full-size, but it must be to scale to enable you to work out the dimensions of everything.

Note that planed 2in × 4in (51mm × 102mm) lumber is actually about 1⁹⁄₁₆in × 3½in (40mm × 89mm), and that other sizes are proportionately smaller.

Whether you are adapting the gate shown in Fig. 1 or designing one of your own, some dimensions will be the same in all cases. These include the sizes of the mortise-and-tenon joints and of the rabbets into which the tongue-and-groove boards are set.

Once you have everything planned to your satisfaction, cut the lumber to length and also cut twelve small wooden wedges from scrap lumber to fit into the mortise-and-tenon joints.

The mortise-and-tenon joint

This joint is very strong and reasonably simple to make. But it must be made accurately, for it is worse than useless if it is a loose fit. Most mortise-and-tenon joints, including the ones here, are held tight with small wooden wedges.

Start making the gate by cutting the lumber to length for the various pieces allowing ½in (13mm) extra for waste. Lay the two stiles, (the vertical outside pieces) face to face and mark off ¼in (6mm) at one end and then from that mark set out the positions of the mortises. Square these lines around the lumber then add an extra ⅛in (3mm) to the top and bottom of the mortise hole on the outer edge of the stile. This is to widen the mortise to make room for the wedges. The mortises should be one-third the thickness of the lumber and you can mark them by setting a gauge to that measurement and scribing a line from each side of the lumber.

Cut the mortise by first drilling it out

Right, top to bottom: Scrap wood is used to raise the gate to the correct height. The gate post is then drilled using the hinge as a template. The outer gatepost is set upright using a level and the gate is then ready for use.

Garden gates

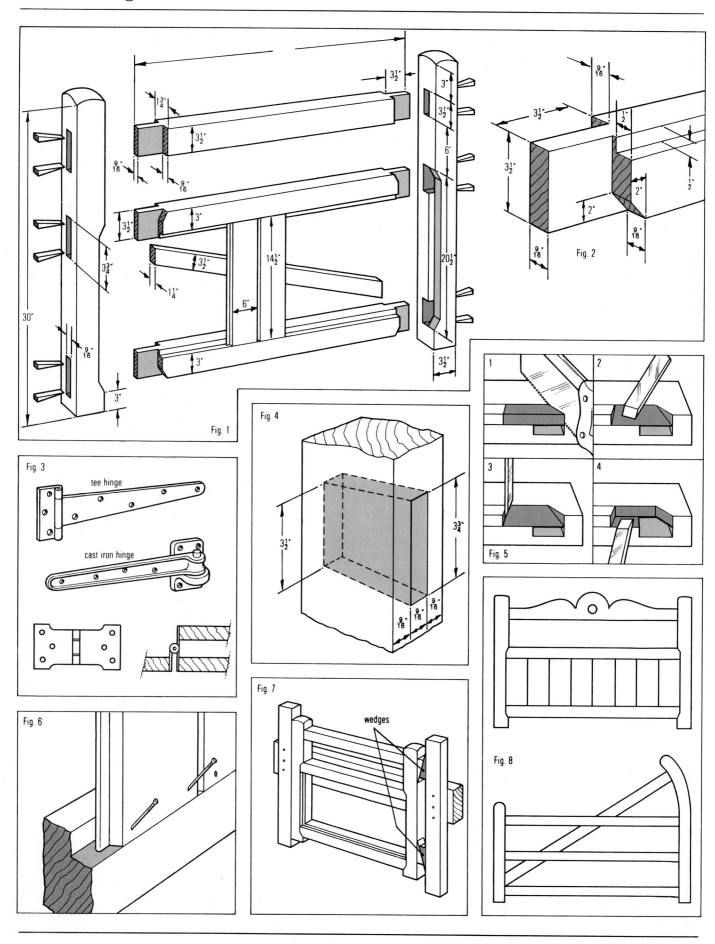

Fig. 1

Fig. 2

Fig. 3

tee hinge

cast iron hinge

Fig. 4

Fig. 5

Fig. 6

Fig. 7

wedges

Fig. 8

making as many holes as close together as you can, using a bit the same size as the mortise hole. The mortise is then squared up with a chisel. Turn the lumber over to square up the mortise from the other side and chisel the extra $\frac{1}{8}$in (3mm) marked on the outer edge of the stile. Taper this off to nothing just inside the inner edge.

Now take the rails and place them together flush at each end, hold them with a small clamp if necessary. Mark $\frac{1}{4}$in (6mm) for waste at one end and then from that mark measure the width of the stiles through which the tenons have to pass. This second line will be the shoulder line of the tenon. To mark the shoulder line at the other end of the rails, take the width of the gate less the width of both the rails and measure this from the shoulder mark you have just made and square a line across the lumber at the other end. You will then be left with the width of the stile plus about $\frac{1}{4}$in (6mm) waste.

Separate the members and square the shoulder lines around each piece. Take the gauge that you used for the mortises and scribe a line along the edge of the rails from the shoulder line right over the end-grain and back to the shoulder line on the other side of the lumber. Do this from both faces of the rail and at both ends. Take the top trail and place it upright in a vise so that you can saw down the waste side of the tenon using a tenon saw. Then cut the waste off by sawing across the shoulder line. Do not cut the middle or bottom rail tenons yet as they have to be adjusted to fit into the rabbet which will be cut in the stiles for the close boarding at the bottom of the gate.

Stopped rabbets

The front inside corners of the stiles between the middle and the bottom mortise holes are rabbeted to take the tongue-and-grooved boarding. The bottom front corner of the middle rail and the top front corner of the bottom rail are also rabbeted. These rabbets are $\frac{1}{2}$in x $\frac{1}{2}$in (13mm x 13mm).

The ends of the rabbets on the stiles are mitered—cut at an angle of 45°—so that

Figs 1–8. The basic construction details and designs of garden gates. The mortise and tenons are unusual because they have to be modified to accommodate the rabbets, required in the stile, to accept the tongue and groove boarding. This should, however, present no problem.

the rails can fit into them neatly. Mark out the full extent of the rabbets accurately, using a bevel for the miter, and make a short tenon saw cut up each end along the line of the miter (see Fig. 5). This cut will not go to the full depth of the rabbet because it only cuts across the corner of the wood, but it gives you a starting point for chiseling out the rabbet end. Make a similar cut 4in (102mm) further along the rabbet, at both ends or however far you need to go to clear a space for the body of the plane. Then chisel out the wood between each pair of cuts, working slowly and accurately to keep the rabbet straight and level. As wood comes away from the ends of the rabbet, chop down deeper past the saw cuts, with a chisel held vertically to expose more end surface. When all the wood is chiseled out, finish the rabbets with the plough plane.

Finishing the parts

When the rabbet is finished in all the members you should check its depth as the tenon shoulder will have to be adjusted to fit into it. On the side of the middle rail and bottom rail on which you have made the rabbets measure $\frac{1}{2}$in (13mm) (the depth of the rabbet) from the existing shoulder line towards the end of the tenon, thus shortening the tenon by $\frac{1}{2}$in (13mm). Now put the rails in a vise one at a time and saw down the waste side of the tenon lines. Take care when cutting the shoulders that you only cut as far as the new line on the rabbet side. It is a good idea to shade the waste on that side of the tenon so that it will remind you.

Having cut the tenons you will have to make a small miter cut on the long shoulder to match the miter on the end of the stile rabbet. Chisel these miters carefully taking off a little at a time until the miters fit neatly. Make a trial fitting of all the joints and make the gate up dry. When you are satisfied with the fit of all the joints, paint all the parts after first giving the knots a coating of shellac or commercial knotting.

You can then glue the joints and clamp the gate up driving the wedges home in the mortises. If you have no bar clamps and cannot hire any, you can make a clamp by fixing two battens to the ends of a length of lumber so that they will span the gate but leaving enough space to insert pairs of wedges (Fig. 7). When you clamp up the gate measure the diagonals to ensure that the frame is square.

Then carefully measure the distance between the rabbeted front edges of the two lower rails. Cut your tongue-and-groove boards slightly over this measurement, and plane them to the exact length. This should ensure that they fit neatly into their rabbets without a gap.

The boarded area on the front of the gate is unlikely to be made up of an exact number of board widths. To give the gate a symmetrical appearance, the edges should be planed off both outside boards to bring the boarding down to width. This process also removes the tongues and grooves, which are not needed because the outside edges of the boarding fit into the plain rabbets on the stiles. The boards must not be too tight a fit; you need to allow for expansion in wet weather.

The only job that remains in the construction of the gate is to make the diagonal brace. This has to be made after the rest of the gate has been completed, because its exact length controls the squareness of the frame. Saw and plane its ends to the right length and angle by measuring them against the frame itself. The brace does not need a mortise-and-tenon joint to hold its place, because it is compressed when the gate is hung. But you must put it in the right way around, with its lower end next to the 'hanging' stile to which the hinges are attached.

Form a slightly domed top on the two stiles (and the gateposts, if you are replacing them) to stop rainwater from collecting on them and rotting the wood. This is best done with a surform, spokeshave or similar tool, working from the edge of the post to the center to avoid snagging the tool on the grain. Also, saw off the ends of the tenons which protrude trough the mortises.

Install the tongue-and-grooved boarding in the rabbets. To avoid splitting the ends of the boards, use oval nails set end-on with the grain of the wood, and drive the nails into the rabbets in the rails at an angle, as shown in Fig. 6. Use two nails for each end of each board if the gate is to be painted, but only one for an unpainted hardwood gate, to allow the boards to widen in wet weather.

When the glue is dry, take the gate out of the clamps and turn it over. Put the diagonal brace in from the back, making sure that it is the right way around. Glue it to the frame at each end, hold it in place with two temporary nails, and then turn the gate front upward again and nail all

Garden gates

the boards to the brace. This keeps both the brace and the boards from warping. No other form of fixing is necessary.

When you have completed the prime coat of paint, screw the hinges to the gate and place it between the gateposts. Raise the gate on blocks or bricks to the height required, then get a helper to hold it steady while you screw the other part of each hinge to the gatepost.

If you are installing new gateposts, they should preferably be set 2ft (610mm) deep in concrete, and at least a week should be allowed (more in cold weather) before putting any weight on them. Posts set in hardcore alone tend to sag unless the hardcore is rammed down with considerable force.

The only two things that remain to be done are to install a latch on the gate and give it its final coats of paint. After that, the gate is complete.

Other gates

If you find the design of the gate given here too plain, there are many variation on the basic gate that do not add too much to the difficulty of building it. For example, for the top rail, you can substitute a piece of 2in x 6in (51mm x 152mm) lumber cut to any shape with a jigsaw (Fig. 8). The boarding can also be cut into decorative shapes.

The farm-type gate also shown is made in the same way as the basic gate, but the diagonal brace runs the other way, so that the weight of the gate stretches it. As a result, the brace must be firmly anchored with mortise-and-tenon joints at each end, strengthened by screws. This type of gate looks best without boarding, so the brace should be as thick as the rails and stiles. Angled halving joints should be made where the brace crosses the rails—it is best to make these while you are actually assembling the frame, to ensure that the angle is correct. The curved stile to which the brace is attached is cut out of piece of 2in x 8in (51mm x 203mm) lumber, using a jigsaw. If you can find a board with a grain that naturally follows the curve you are going to cut, the stile will be much stronger.

If you are prepared to be more adventurous with techniques, the only limit on the designs you can build will be set by your imagination.

Above: A farm gate with diagonal bracing.

Top and right: A pair of simple wicket gates and a creosoted farm-type gate suitable for a wide drive.

Cutting list

Solid wood	standard	metric
Frame	2 x 4 x 240	51 x 102 x 6096
Diagonal brace	1½ x 4 x 48	38 x 102 x 1219
T&G boarding	½ x 6 x 132	13 x 152 x 3352
Gateposts	3 x 3 x 120	76 x 76 x 3048
Gateposts (alternative)	4 x 4 x 120	102 x 102 x 3048

You will also require:

Knotting. Primer. Undercoat. Paint. Galvanized oval head screws 1½in (38mm). Black japanned screws. Hinges.

Playhouse

Playhouse

Left: Details of the playhouse supports. These are the most important parts to consider in the construction as the stability of the frame structure depends upon them.

Fig 1. The frames for the playhouse are built to this pattern.

Fig 2. View of the front and side elevation of the playhouse showing the guys and rails.

Construction

The basic construction of the playhouse is quite simple. It consists of two lumber triangles, stood on their apex and bolted together through the short side of the triangles—the top member in the finished construction. The apex of each triangle is fixed to a concrete footing, set into the ground. The structure is braced by metal cables, fixed to the top member of the triangles and running to the ground, where they are attached to hooks set in concrete footings.

A look-out platform is placed between the wood members and the metal guy ropes, about halfway up the height of the playhouse. One end of the platform has a rectangular opening—this acts as a kind of trap door to the platform. At the opposite end there is a triangular wall, with an opening in the base of it which acts as another door. Nylon ropes are strung around the playhouse members and the metal guys. These act as safety rails around the platform.

The platform is reached by means of 'rope ladders'—in this case made from nylon rope with wooden rungs. These are securely fixed at the top and bottom.

Planning considerations

The playhouse will occupy an area in your garden measuring 9ft × 9ft 8in (2.7m × 2.9m). It is worth drawing a scaled plan of your garden to help you position the playhouse correctly—it can be moved after it has been erected, but planning will save you the trouble.

The best position for the construction is near to the house where you can keep an eye on the children. Do not build the playhouse near an out building or tall trees—children are likely to use the playhouse to climb into higher, and more dangerous, places. Build the playhouse on a lawn so that if your children do fall out of it they will land on a fairly soft surface.

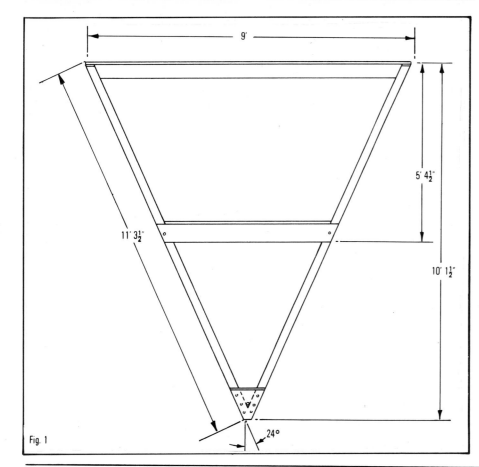

Fig. 1

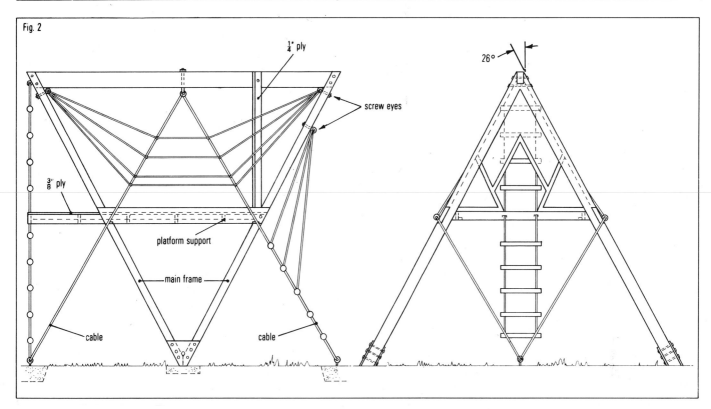

Fig. 2

¼" ply

screw eyes

26°

⅜" ply

platform support

main frame

cable

cable

The 'A' frames

The 'A' frames are made from lengths of 2in x 4in (51mm x 102mm) lumber to the shape shown in Fig. 1. The lumber that forms the long sides of the triangle is joined to that which forms the short side with housing joints. At the apex of the triangle the two long sides are butted. These joints are strengthened by plywood gussets bolted in place over the apex of the triangle on both sides of the lumber.

Cut the long sides of the 'A' a little overlong using a circular saw and combination blade. On the ends of these members, mark an angle of 156° on their narrow edge. You can mark this with a bevel gauge, setting the angle with a protractor. Cut through the lumber, down the marked lines.

Now mark the finished length of the members, referring to Fig. 1 for dimensions. From the point marking the finished length, mark an angle of 156° on the wide edge of the members. Cut through the lumber, down the marked lines. Next, cut, in the newly cut slanted ends, a 1in (25mm) housing to take the ends of the short sides of the 'A'.

Cut the ½in (13mm) plywood gussets that cover the apex of the 'A' or triangle. These gussets are shown in Fig. 1. Bolt these in place over the apex of the triangle with 5in (127mm) carriage bolts. You now have two V-shaped constructions. Cut the short sides of the triangles oversize and lay them between the arms of the V, set into the housing. Mark the length of this side of the 'A'. Cut it to length and nail it to the two arms of the V-shaped constructions. This gives you the two triangular frames.

The next step is to cut housings in the frame members for the two members that form the side struts of the look-out platform. The position and all the necessary dimensions of these are shown in Fig. 1. Nail these side struts in place in the housings.

Now bolt the two triangles together through the short sides. Use 7in (178mm) carriage bolts for this, spaced about 12in (305mm) apart along the lumber. At the end use two carriage bolts, one underneath the other. The top one should be 7in (178mm) long, the bottom one 10in (254mm) long.

The concrete bases

With the two triangles bolted together, you can now determine the exact position for the concrete bases for the legs and guy ropes. To do this you will have to hold the construction up in its finished position—you will need at least two helpers for this.

With the helpers holding the triangles upright, mark around the point on the ground where the legs of the playhouse rest. The guy ropes run to points on the ground almost underneath the ends of the short sides of the triangles, see Fig. 2. You can estimate the position of the blocks here or mark out the ground plan with wooden pegs.

Once you have determined the position of the bases you can dig the holes for the concrete footings. These should be about 24in (610mm) deep. The holes for the guy rope blocks should be wider at the bottom than at the top—the strain exerted here is an upward pull, whereas the pressure exerted by the legs is downward—the blocks for these can, therefore be straightsided. All four blocks stand proud of the ground by 4in (102mm) so you will need a simple former. One of the concrete footings is shown in Fig. 2.

As a fixing for the guy ropes, a ¾in x 6in (19mm x 152mm) eye bolt with a 1in (25mm) diameter eye on the top is set into the block. Or you can make these yourself from lengths of ⅜in x 10in (10mm x 254mm) mild steel rod. They are bedded in the concrete so that just the eye stands proud. The fixing for the wood legs of the playhouse is a 1½in x 12in (38mm x 305mm) diameter pipe, bedded in concrete to a depth of 6in (152mm). A flange joint is slipped over the pipe so that it sits on top of the block.

Pour the concrete for the blocks and push the fixings into it. Leave the concrete to set.

Playhouse

The look-out platform

This consists of a framework of 2in x 3in (51mm x 76mm) dressed 4 sides lumber boarded over with ⅜in (10mm) exterior fir AC plywood. The platform is bolted to the inside edge of the long sides of the triangles, at the points where the side struts of the platform wall are housed into the triangular frame members. The platform has a rectangular opening at one end to act as a door to the platform. Construct the platform to the shape and dimensions shown in Fig. 3.

To bolt the square-sided platform to the sloping sides of the triangles, you will first have to cut four wedges. Lightly nail these in place on the triangle members. The platform is bolted in place with ⅜in x 8in (10mm x 203mm) eye bolts. These also give a fixing for the guy ropes that are positioned later. Bolt the platform in place.

Use ⅜in (10mm) exterior fir AC plywood for the side wall of the look-out platform. This is nailed to the platform and to the side struts fixed earlier. Before you do this though, cut an opening at the base of the panel to be used for the side wall—this acts as another door. The shape of the opening and the position of the extra lumber that frame it are shown in Fig. 2.

The rigging

The rigging that runs to the two concrete blocks set in the ground is 3/16 in (5mm) nylon tiller rope. This has a 100lb (450kg) breaking point and would be very difficult to cut through.

You will need four lengths of cable, each about 11ft (3.3m) long. These run from an eye bolt and ring fixed through the top member of the two triangles, in the center of their length.

They then pass through the eye bolts on the sides of the look-out platform and around the triangle member to the concrete blocks. At the top the cable itself is attached to ring bolts with shackles, and at the bottom to twin buckles with cable clamps.

Drill a 1½in (38mm) hole up through the bottom of the wood legs. Do this with a brace and expansive bit. These holes allow the legs to be fitted over the metal pipes set in the concrete. Drill a ½in (13mm) hole in the bottom of the plywood gussets also—this will allow rain-water to drain out of the base of the legs.

The safety rail

This consists of nylon rope, strung from the ends of the top member of the triangles and fanning out to plastic fittings on the guy ropes. The arrangement of the nylon rope is shown in Fig. 2.

Erecting the playhouse

Again you will need two helpers for this. First, fix the guy ropes to the fixing in the top triangle member and thread them through the eyes of the fence stretchers in the platform sides. Get your helpers to lift the playhouse in place, with the wood legs over the metal pipe fittings. Attach the twinbuckle on the end of the guy ropes to the eye fittings in the concrete footings. Tighten the guy ropes.

The rope ladders

The rope ladders are made from 12in (305mm) lengths of 1⅝in (41mm) closet pole strung onto nylon ropes. Drill ½in (13mm) holes through the members, about 1in (25mm) from their ends. Take a length of nylon rope. Tie a knot in it. Do the same with another length of rope. Repeat this process, spacing the rungs every 9in (229mm) apart until the ladders are complete.

Both ladders are fixed at the bottom to the guy ropes, 9in (229mm) from the ground. Fix them to the guys with the plastic eye clips used to fix nylon safety rail to the guys above the platform, or use rope thimbles and splice the rope around them.

One ladder is vertical and runs through the opening in the platform floor. At the top, about 12in (305mm) from the top triangle member, the two lengths of rope are tied to size 14 screw eyes. The other ladder runs up to the opening in the platform wall. It is tied to the side of the platform, through size 14 screw eyes.

Finishing

You can coat all the members with a wood preservative, or paint them in the color of your choice. This will weatherproof the lumber. Decoration of the playhouse—decking the guy ropes with colored pennants, for example—is up to you.

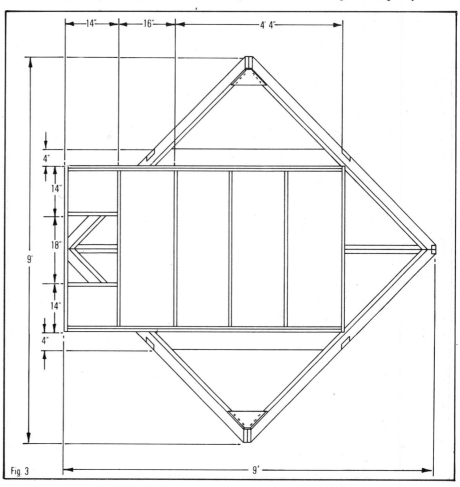

Fig 3. The completed playhouse as it is viewed from below.

Fig. 3

Jungle gym

The jungle gym shown here consists of an easily-constructed climbing frame and slide.

This play area has been designed to that it can be altered to suit the dimensions of any particular garden. It consists of a climbing frame, slide and sand pit, any of which could be scaled down in size or omitted if you do not have enough room.

The main feature is the climbing frame, and as this will be used by children of differing ages, platforms of varying heights have been incorporated. A slide has been included because it is one of the most popular playground items. And any timid types who do not wish to climb the rungs can make use of the stepped ladder which hooks on to one of the platforms.

The frame consists of towers made up from 2½in (64mm) (have cut to order) square uprights in modules of 2ft 4in (711mm), the highest towers being 6ft (1.8m), and the lowest platform 2ft (610mm). The towers are joined by 1in x 2½in (25mm x 64mm) cross members and dowel rods 1⅛in (29mm) in diameter. These also provide the climbing rungs.

Jungle gym

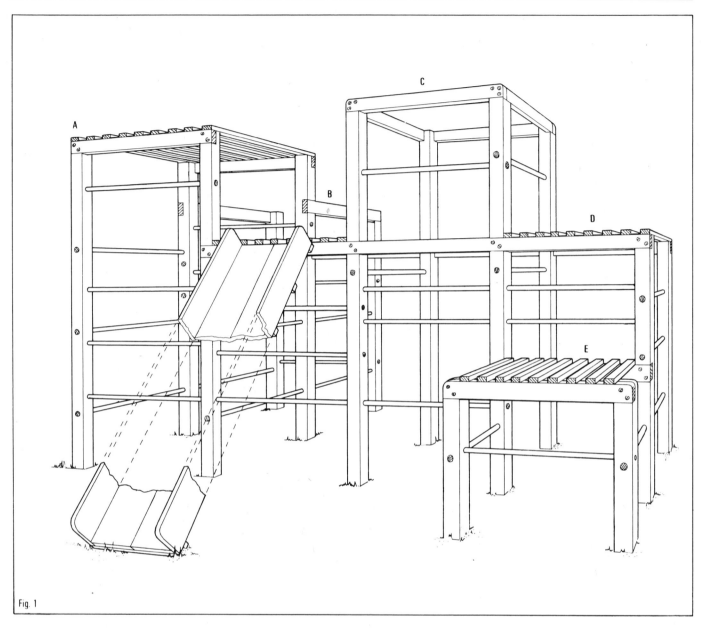

Fig. 1

Planning the project

It is not essential to follow the exact layout of the frame described here, as long as the structure is stable and strong. But various factors should be taken into account when planning the unit.

Safety is the main consideration. As the frame is likely to be used by several children together, a tower must be placed at the rear, and another at the front of the frame. These act as stabilizers, preventing the frame from toppling over if all the children happen to be playing on one section of it.

Your garden may not have enough level space for the design shown here. In this case you will have to limit your frame, or vary the tower legs to allow for irregularities or sloping ground

Hardwood is an expensive item and this, too, may limit the final size.

At the planning stage you should consider the placing of the feet of the frame By placing them on concrete slabs you will be able to keep the grass from growing untidily against the uprights.

If you are altering the design of the structure, do not try to eliminate the doweling that spans two modules. Doweling that is inserted through three uprights makes for a stronger structure than doweling through two uprights. Also, make sure the bottom rung is high enough to provide clearance for a lawn mower, as in the original frame described here. Otherwise you will have an untidy patch inside.

Materials

Materials for the climbing frame are in the cutting list. The eight 6ft (1.8m) lengths are for towers A and C, and the two 5ft (1.5m) lengths are for tower B, which is a stabilizing structure. This is joined to tower A by bearers. The two 4ft (1.2m) lengths are for the two legs at the right

Fig 1. An illustration of the completed climbing frame and slide that make up the jungle gym. As you can see, the construction of this unit is extremely simple and should present no problems.

Fig 2. Uprights and joints for the cross-rails are cut to the dimensions shown. It is important to make sure the joints are cut on the correct side of the lumber.

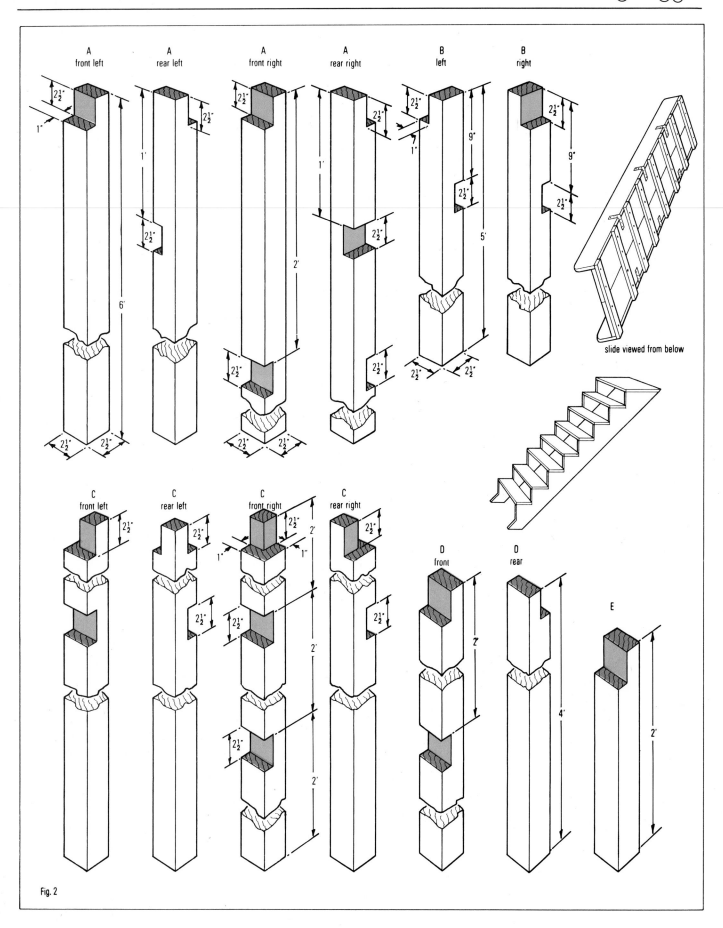

A
front left

A
rear left

A
front right

A
rear right

B
left

B
right

slide viewed from below

C
front left

C
rear left

C
front right

C
rear right

D
front

D
rear

E

Fig. 2

end (D) of the frame, and the shortest lengths are for the lowest jumping platform, E, which acts as a front stabilizer.

Tower B is a stabilizing structure at the rear of the frame. The two 5ft (1.5m) uprights have joints cut at the top of each to take the bearers which connect it to tower A. Additional joints 9in (229mm) down are cut to take a cross member.

Tower C is a hollow climbing frame (it has no platform on the top). It consists of four uprights, 6ft (1.8m) in length, held together at the top by bearers.

The addition of tower D is made by two 4ft (1.2m) uprights, which join to the main cross members linking towers A and C. These members also act as bearers for the platform battening on the top of tower D.

The front of the structure is stabilized by tower E, the low level jumping platform. This consists of two 2ft (610mm) lengths which act as uprights, joined by a bearer on the front, and at the back to the bearer on the front of tower D.

Cross members and bearers

These are used to connect the three main towers as well as to strengthen the tower tops. (Cross members are horizontal and used to strenghen a structure: bearers both strengthen and support a load—in this case, the platform battens.)

For safety alone it is important that all cross members fit snugly at the joints and are securely screwed and glued. Several children weigh a considerable amount.

Constructing the frame

Plane all lumber to a smooth finish to prevent splinters from forming. Sand, then mark out the joints as indicated in the diagrams. Each upright is described as though looking at the front of the front of the frame—that is, the face to which the slide is attached.

The joints are all of the simple halving type, which makes the unit particularly easy to assemble. But take care to mark the joints on the correct faces. It will be a great help if you pencil an identification on each upright. Countersink all screws, using a countersinking bore in your drill.

The two main bearers are the 6ft 3in (1.9m) lengths at the front and rear. Tap these into the joints of the uprights, drill holes for screws, then screw and glue to the front uprights. Assemble the rear uprights and tower B in the same

manner. This allows the top bearers to be screwed and glue into place, and the battens to be screwed in place between towers A and C to form the platform access between ladder and slide. Allow a gap of 1in (25mm) between the battens. Then screw the battens for towers D and E in place, and fix the top of tower C with the four bearers two of 2ft 4in (711mm), two of 2ft 2in (660mm).

Fitting the rungs

Once the main frame is screwed together, the next step is to mark out the centers for boring the holes for the dowelling.

It is essential to work out with care the spacing of the dowelling. The rungs should not be too close to one another, as this takes the fun out of climbing. Also, where two dowels are to be inserted through an upright in different directions (at right angles) the holes should not be bored too close to each other as the upright will be weakened and might split.

First, mark out the center line on the uprights with a try square. Then mark a line halfway at right angles to produce a cross, the middle of which will be drilled for the dowel.

Secure the dowelling jig so that the guide hole is right over the mark, then drill the holes with a No. 8 bit on the power drill.

When you have drilled all the small No. 8 holes, thread a string line through them. This will indicate whether the line is running correctly, at right angles to the uprights.

Now insert in the brace a bit the same size as the dowelling, and bore out the holes. Do this fairly slowly. As soon as the point of the bit shows through the pilot hole, stop and drill from the opposite direction, using the pilot hole for centering. This will prevent the edges of the hole from splintering.

Now clean up around the inside of each hole with a Surform or curved wood rasp.

You are now ready to tap the dowels through with the mallet. It helps to have someone to give some support to the uprights when knocking in the first few dowels. Any spare dowelling which sticks out should be left until the frame is well and truly squared up. Then the spare bits can be cut away.

Making the slide

Place the two 7in (179mm) planks side by side and secure them with the eight battens. Allow the battens to overlap the edges of the planks by 1in (25mm) on both sides.

To form the sides of the slide, place the 4in (102mm) planks on the overlapping battens so that they are at right angles with the slide planks. First screw the sides onto the edges of the slide, then screw the battens to the bottom of the side planks.

Secure the three steel L-brakets down each side for additional support.

Plane any corner, on the top of the sides and the top and bottom of the side itself—to half round, so that no nasty angles protrude.

The method of attaching the slide to the fame depends on whether you want it to be screwed in a semi-permanent manner or hooked on so that it can be detached quickly. An L-bracket could easily be bent to fit, and screwed to the underside of the slide, while a simple hook-and-eye provides an easy slip-on fitting.

Cutting the ladder steps

The ladder is secured to the side of the tower opposite the slide. It is made out of two 4ft 6in (1.3m) lengths of lumber with shorter planks providing the steps.

Recesses for the steps are cut out by sawing uniform V sections along one side of each plank. This operation requires nothing more than a measuring rule, a pencil and saw.

To mark out the step positions, begin by fixing one side of the ladder temporarily in place to establish the correct slope. Next, use a builder's level and pencil to mark one horizontal guide line across the board. (This is to ensure that the steps are level.) Remove the ladder side and lay it flat while you do the rest of the marking out. For this, you will need a cardboard or hardboard triangle, whose shortest side is 3in (76mm) and whose longest side matches the slope of the steps. Simply by sliding the triangle along the board you can mark each successive step on the face side. Use the try square to carry the marks across the edge of the board, and the cardboard triangle to mark the other side. Finally, mark out the second board so that it matches the first.

Saw along the zig-zag lines to cut out the outlines as in Fig. F.

Screw the steps in position and round the edges with the plane.

The ladder may be attached to the frame in the same way as the slide.

Polyurethane varnish should now be applied to all woodwork. In view of the fact that this unit will spend its life outdoors, two coats are recommended.

The sandpit

The area of this sandpit is approximately 4ft x 6ft (1.2m x 1.8m).

Dig the pit to a depth of 2ft 6in (762mm) and line it with any old planking that you can get. Creosote the planking well. Secure the planks at the corners by simple square stakes driven into the ground, and then back-fill them. When the top edge of the planking is level—a visual sighting is sufficient for this—nail through the stakes into the planks.

For this pit ornamental square concrete slabs are laid around the edge, giving a slight overhang which provides an on-site seating arrangement, and a firm base for building sand castles. The slabs also make it easier to sweep the surround free of excavated sand. If possible, obtain slabs with rounded edges—these are kinder to children's legs.

To assist drainage, line the bottom 6in (152mm) of the pit with gravel and ram it down well. Finally, fill the pit with sand (of the non-staining variety, if possible).

Cutting list

Solid wood	standard	metric
Lumber for uprights		
8 lengths	$2\frac{1}{2}$ x $2\frac{1}{2}$ x 72	64 x 64 x 1829
2 lengths	$2\frac{1}{2}$ x $2\frac{1}{2}$ x 60	64 x 64 x 1524
2 lengths	$2\frac{1}{2}$ x $2\frac{1}{2}$ x 48	64 x 64 x 1250
2 lengths	$2\frac{1}{2}$ x $2\frac{1}{2}$ x 24	64 x 64 x 600
Lumber for cross members and bearers		
2 lengths	1 x $2\frac{1}{2}$ x 75	25 x 64 x 1905
9 lengths	1 x $2\frac{1}{2}$ x 28	25 x 64 x 711
2 lengths	1 x $2\frac{1}{2}$ x 26	25 x 64 x 660
Lumber for platform battens		
37 lengths	1 x 2 x 28	25 x 51 x 711
Hardwood dowelling for rungs		
11 lengths	$1\frac{1}{8}$ dia. x $28\frac{1}{2}$	28 dia. x 724
23 lengths	$1\frac{1}{8}$ dia. x 54	28 dia. x 1372
Lumber for slide (pine planks)		
2 lengths	1 x 7 x 96	25 x 178 x 2438
2 lengths	1 x 48 x 48	25 x 1219 x 1219
Battening for underside of slide		
8 lengths	1 x 2 x 16	25 x 51 x 406
Lumber for ladder		
2 lengths	1 x $4\frac{1}{2}$ x 54	25 x 112 x 1372
8 lengths	$\frac{3}{4}$ x 3 x 12	19 x 76 x 305

You will also require:

A quantity of No.8 screws 2in (51mm) long; some of $1\frac{1}{2}$in (38mm); some $\frac{3}{4}$in (19mm). Nails. 6 steel L brackets 3in (76mm) max. Steel strips or hooks. Wood adhesive.

Gazebo

The 'gazebo', or octagonal greenhouse, can be built largely of softwood or entirely of hardwood. The lower 'cheeks' of the greenhouse sides are lumber clad; but plastic could be used here as well.

The greenhouse is largely prefabricated, then assembled on its foundations. The main structure consists of eight equal panels, forming the octagonal shape when assembled.

Site preparation

Select a position for the greenhouse which makes the best use of sunlight, yet gives ready access. Next, level the overall site area, using pegs, a straightedge and a level. It is a good idea to provide a slight incline in the level of the ground so that water does not collect around the base of the greenhouse.

The footings are then marked out. Find the overall depth of the structure, from front to back (in the case of the greenhouse illustrated it is 5ft [1.5m]), and mark this, using a measured string line. Place a marked peg at each end of the line. Find the center position and then take the string across to form a cross; place pegs at each end. Measure between the four pegs to find the halfway points and put in four more pegs, the same distance as before from the center position, but this time in the form of a letter X. This gives a roughly octagonal shape (Fig. 1). Now mark at right angles with a spade, at the top and bottom and the side pegs of the cross, the width of the frames which form the greenhouse sides. Join up these marks diagonally. This gives an accurate octagonal profile.

Mark out with the spade the line of the footings. Do this by marking slightly in front of the octagonal outline, and also slightly deeper than the greenhouse side frames, giving a total width of about 6in (152mm). The foundations need to be slightly wider overall than the structure to provide a firm bearing.

Excavate a hole to a depth that goes below the frost line and place lumber forms in this, above the ground by about 3in (76mm). It is important to raise the foundations slightly above the surrounding earth, since this could rot the greenhouse wood.

The form should be level, since the top of this is the top of the concrete. Use a straightedge and level to check the form levels.

Once the holes are filled with concrete, this is well tamped down and allowed to set. Forms are then removed, leaving the footings with an upstand.

Finally, lay 4½in (114mm) 15lb (6.75 kg) felt on top of the footings, to prevent damp rising into the wood framework.

As an alternative to concrete footings, the greenhouse can stand on raised paving or brick. The main requirements are that these footings are both firm and level, otherwise, it will not be possible to assemble the greenhouse accurately on site.

Lumber preparation

Depending on the type of lumber you use, and the climate in which you are building, some pre-treatment of the lumber may be necessary to prevent its rotting quickly. In some circumstances, every joint in the lumber—particularly end-grain—must be given a thick coat of prime paint as the joints are assembled. When you buy your lumber ask your lumber dealer for advice on this point.

Building the wall frames

The first stage is to make up the four opposed square frames (Fig. 2). Frames A and B are made in the same way up, except that the former uses a doorstep section in place of the bottom rail (Fig. 3). Frames are made from 2in x 4in (51mm x 102mm) dressed lumber, jointed together by 3in (76mm) 10d nails inside the uprights of each frame. For all construction use non-rusting galvanized or aluminum nails.

Half-lap joints are cut with a circular saw at an angle of 45° in the top rail sections before assembly. Nails are then driven through these joints into the top ends of the side stiles (Fig. 4). Details of the joints are shown in Fig. 5.

Sections marked 'triangle X' on Fig. 4 are sawn from 4in x 4in (102mm x 102mm) softwood to the desired height of the windowsills. This is simply a matter of deciding how much each window area above these you want.

Next cut the base rails (shoes) C and place a piece of section X at the end and draw the angle. The V section is formed by returning the angle at 45°. (A piece of section X can be used to help scribe this.) The wedge is then cut out as shown in Fig. 7. This procedure is followed at both ends of rail C.

The X and C sections are nailed together as shown in Fig. 6. In Fig. 8, B, C and X are seen butted together.

The completed framework is assembled on the damp-proof felt (Fig. 3) and joined together by nailing through the triangle X sections into adjacent uprights of frames A and B. Additional support can be given by fixing angle brackets to the internal edges of door rails B and C and doorstop A.

The top rails C1 (Fig. 9) are angle cut to 45° and lapped. These are slotted into the lapped sections of frames A and B and nailed downward into the side rails.

Angle brackets can be screwed to the internal side of the top rails to provide added strength.

Cladding and windowsills

Horizontal siding (Fig. 10). Unless you have a fair amount of experience with this material, it is tricky to work with. The boards must be mitered at 22½° at both ends, because cutting one board to 90° and the adjoining one to 45° would throw the vertical joints off the true line of the corner. Two things will help you get a snug fit: (1) cut the miters just a fraction too 'sharp', so that no gaps show at the outside edges when they are joined and (2) as you fix each row of boards, butt them against a temporary vertical stop (mitered lengthwise at an angle of 22½°) so that the ends of all the boards align accurately with one another.

Vertical boarding is much simpler, since only the boards at each end of any particular wall need to be mitered—that is, you make two miter cuts per wall, instead of a dozen or more. You will need a sub-sill to which to fasten the top ends of the boards. For C sections cut these from 2in x 4in (51mm x 102mm) lumber the same shape as the base rails. For B sections, cut them square on both ends the length of side T&G on two 2in x 4in (51mm x 102mm) lumbers. Start by vertically mitering a pair of boards a 'hair' sharper than 22½°, and fix them around the angle farthest from the door. Then work progressively back toward the door. Whichever type of board you use, fix it with non-rusting rails.

Once all the siding is in place, the sills can be fixed. Try to buy the type which has a flat base.

Each length of sill must be notched at the back (Fig. 11) to accept the vertical members of the greenhouse frame. At its outside edge, it must be cut accurately to 22½° to meet the adjoining length of sill. If, when fitting the sills, two adjoining pieces are found to be slightly overlong, this can be adjusted by running a saw

Gazebo

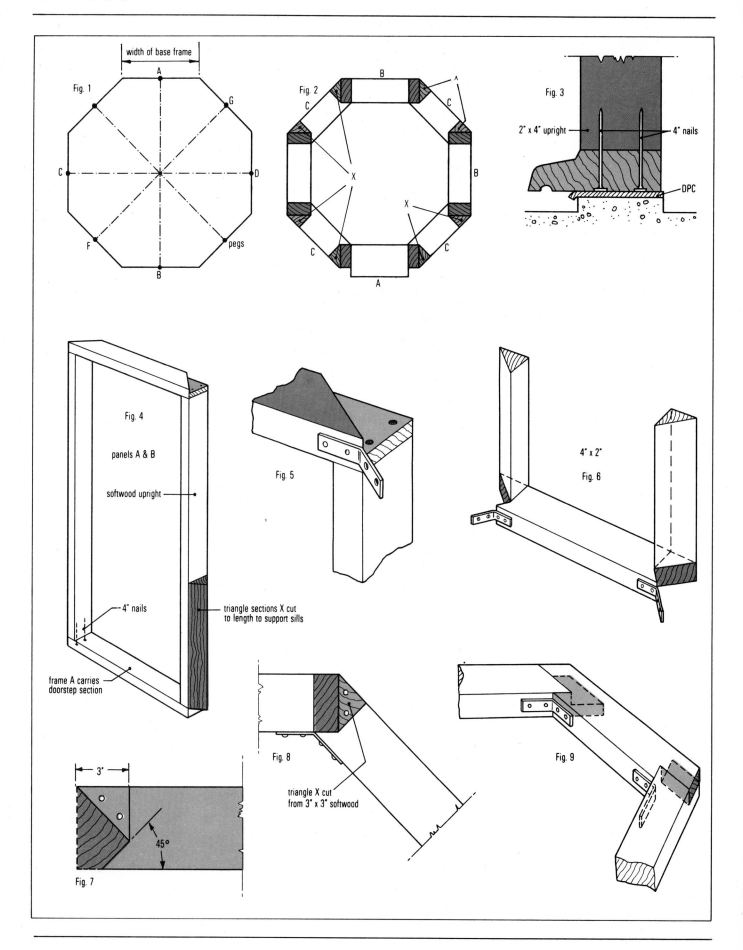

width of base frame

Fig. 1

A

G

C

D

F

B

pegs

Fig. 2

B

C

C

X

B

X

C

C

A

Fig. 3

2" x 4" upright

4" nails

DPC

Fig. 4

panels A & B

softwood upright

4" nails

frame A carries
doorstep section

triangle sections X cut
to length to support sills

Fig. 5

4" x 2"

Fig. 6

3"

45°

Fig. 7

Fig. 8

triangle X cut
from 3" x 3" softwood

Fig. 9

Fig 1. Ground plan of the gazebo. Pegs are used to mark out the site before cutting the trench for the foundations.

Fig 2. Plan view of the gazebo. Note the doorstep section which replaces the bottom of one of the frames A.

Fig 3. Section through the frame A showing the doorstep.

Fig 4. Construction of the frames A. The pieces X are sawn from 4in x 4in (102mm x 102mm) softwood.

Fig 5. Details of the top corners of frames A showing the angle iron used to strengthen the joint.

Fig 6. The lower edge of the frames showing the joint between pieces C and X.

Fig 7. Detail of the corner joint between pieces C and X. A piece of triangular section lumber is used to scribe the cutting line on piece C.

Fig 8. The completed joint. Note the angle iron.

Fig 9. Detail of the top end of the construction showing piece C1 and the method used to join it to the frames A.

cut through the joint—but be careful not to score the paneling beneath.

The roof rafters W are cut from eight pieces of 2in x 4in (51mm x 102mm) softwood and pitched at an angle of 10° to converge at a center point (Fig. 17). The convergent top ends are mitered to $12\frac{1}{2}$°, and trimmed back to make a flat platform of about $5\frac{1}{2}$in (139mm). At the lower end they are cut to a 'birdsmouth' to fit over the top rails A, B and C1— a cardboard template will help get the angles right—and slightly over hang the sides to take rainwater clear. Detail of the roof structure is shown in Fig. 12.

Next, cut eight sections of triangular lumber, Z, diagonally from 1in x 4in (25mm x 102mm) softwood (Fig. 16). These are cut to fit between rafters and are nailed to the top rails of frames A, B and C1, and W. The ends of the sections are mitered to an angle of $22\frac{1}{2}$°, the angle at which the roof struts converge.

An octagonal capping piece is fixed to the roof struts at the point of convergence, with a similar octagonal piece beneath them (Figs. 12 and 13). These are cut from hardwood and screwed into the struts with brass screws to form a water cover to carry away water from the open joint beneath the capping. They also add extra strength to the structure. The pieces are made from 1in x 6in (25mm x 152mm) blocks of hardwood—beech, oak or mahogany. A block plane is used to shape the sections to produce the eight upper faces; the corners are cut away to form an octagon (Fig. 14). The capping piece is bedded on mastic; this is not necessary for the piece on the underside (Fig. 15).

Cut glazing strips $\frac{3}{4}$in x 1in (19mm x 25mm) softwood and nail these to each side of the roof rafter W between the triangular sections Z and the center point of the roof. The top surface edge of S should form a continuous line with the top surface of Z, to provide support for the glass roof.

Eight pieces of glass, cut to the triangular shape of the roof, are required, each to overhang the outer edges of Z by about 1in (25mm). The panels are bedded down on to putty spread over the top edges of S and Z.

Eight sections of hardwood R, are cut to fit to the exact lengths between roof rafters W. Hardwood angle section sould be used here. The top edges should be rabbeted to receive, later, the ends of the glass panels Y. Sections R are nailed and

glued to the roof struts W and screwed through the center of R to top rails B and C1. Bed putty between the glass and rabbet after fixing.

Cut sixteen hardwood fixing Patterns T (Figs. 16 and 17) to the length of rafters W and fix them together to form a flush top surface (Fig. 17). Putty should be bedded between the glass and under the edge of section T.

Outward-opening pivot widows can be fitted, if desired. It is a sensible provision to provide greenhouse ventilation; two or three windows would do.

The windows are made up to size using rabbeted lumber. The choice of joint to make the frames is a matter of personal preference. The frames are hinged to swing open from the top and can be kept open by fitting conventional window stays at the bottom.

Figs 18 and 19 show glazing details. The glass is fixed in position by nailing in place window glass-fixing beads H at the rear. The glass is lightly bedded in putty or mastic.

The fixed lights are similarly bedded and fixed in position by four fixing beads. The lower front one K is angled forward slightly to allow water to run down.

The fixing beads in the angled windows between the main frames are chamfered at an angle of 45° (Fig. 20).

It is possible to make the door, but a suitable white wood one may be picked up 'off the shelf' at a lumberyard. The door should be suitable for full or partial glazing—again, a matter of personal choice.

Fixing the door

The door should be fixed to open outward, to give the maximum space in the greenhouse and permit benches to be fitted inside.

The long rails, or stiles, have protruding ends, known as horns, which you have to saw off. These are to protect the door in transit and storage. The door may have to be planed to fit well, but first it should be tried against the opening.

Planing should start from the edge on which the hinges are to be fitted, known as the hanging stile. You should aim to make this stile as good a fit as possible to any curves or variations in the frame. A jack plane is best here, since this will give a truer edge. If the edge is planed to a slight bevel, this will give slight clearance without increase of the visible gap.

Gazebo

Fig 10. The method of fixing the cladding and windowsills. The ends of the cladding need to be mitered at $22\frac{1}{2}°$. Accuracy is necessary in order to make a good job of this part of the construction.

Fig 11. The windowsills require careful fitting in order to produce a professional appearance.

Fig 12. Details of the roof construction .

Fig 13. The center of the roof of the gazebo showing the patrice and the method of fixing it to the rafters.

Fig 14. The capping piece is marked out as shown and cut to shape using a block plane.

Fig 15. Detail of the center of the roof showing the capping piece, which is bedded in mastic to make the joint watertight.

Fig 16. Detail of the roofing at the lower end of the rafters.

Fig 17. Plan view of the roof of the gazebo showing the method of joining the rafters an pieces T.

Fig 18. Section through a window showing the method of hinging and the position of the surround.

Fig 19. Section through the fixed lights showing the beading which holds the glass.

Fig 20. Section through the vertical frames. Note the shape of the beading above the pieces C between the frames A.

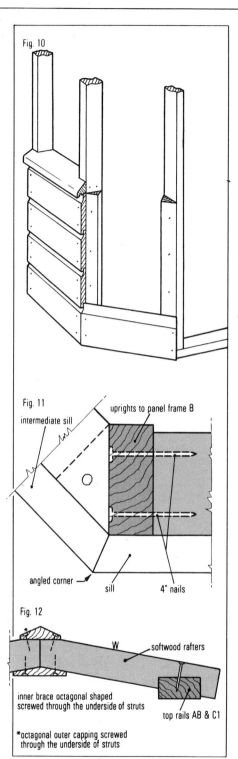

Fig. 10

Fig. 11
intermediate sill
uprights to panel frame B
angled corner
sill
4″ nails

Fig. 12
W
softwood rafters
inner brace octagonal shaped
screwed through the underside of struts
top rails AB & C1
*octagonal outer capping screwed
 through the underside of struts

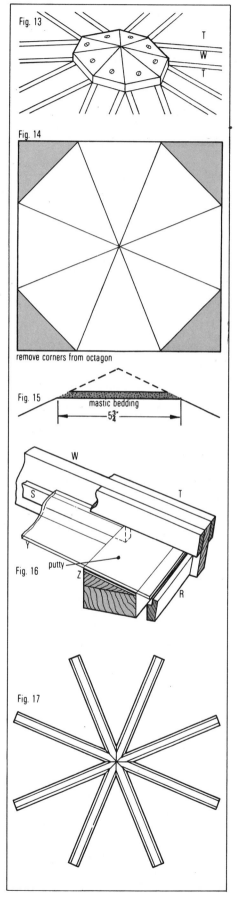

Fig. 13
T
W
T

Fig. 14
remove corners from octagon

Fig. 15
mastic bedding
$5\frac{3}{4}″$

W
S
T
Fig. 16
putty
Z
R

Fig. 17

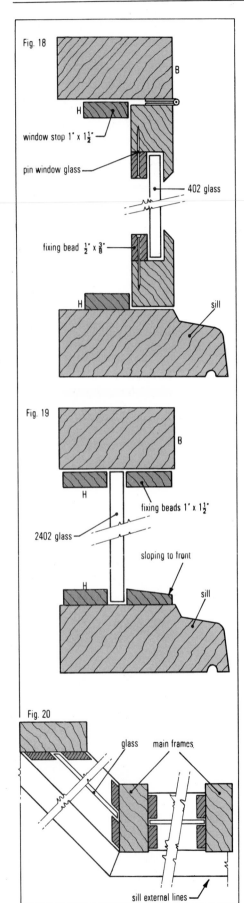

Fig. 18

B

H

window stop 1' x 1½'

pin window glass

402 glass

fixing bead ½' x ⅜'

H

sill

Fig. 19

B

H

fixing beads 1' x 1½'

2402 glass

sloping to front

H

sill

Fig. 20

glass   main frames

sloping to front

sill external lines

Once the hanging stile is fitted accurately to the jamb, plane the opposite stile. This must have a slight bevel, of about $\frac{1}{16}$in (2mm) for a satisfactory fit. A top and edge clearance of about $\frac{3}{32}$in (2.3mm) is necessary to allow for painting.

When trying the head of the door, allow a little less clearance above the lock stile, since doors tend to drop slightly as hinges wear. The bottom rail should have a clearance of about $\frac{3}{16}$in (5mm). Allowance must be made for an aluminum saddle fixed across the bottom of the door.

Hinges and latches

Cast-iron hinges or butts can be used to hang the door, though brass or plastic are suitable. Pressed-steel butts are not as strong as cast ones and may rust, unless you can buy galvanized. The door may be hung on two hinges, though three would spread the work load and prevent possible middle distortion. Use three 3in (76mm) pessed-steel hinges or two $3\frac{1}{2}$in (89mm) or 4in (102mm) cast butts.

The depth of each leaf is marked on both the door and the door frame. Each leaf is recessed to this depth Using a chisel, make a series of cuts across the grain as deep as the gauge line, then pare with the grain to remove the waste (Fig 21). The hinges can now be screwed to the door which is then fitted into the frame opening. Slide a wedge underneath and a piece of $\frac{3}{32}$in (2.3mm) packing at the top to line up the door in position. The edge positions of the hinges can then be penciled onto the frame, squared into this using a try square and marking gauged to leaf depth. The recesses are chopped out in the same way as those of the door.

Try the door by inserting one screw in each hinge. Provided no adjustments are needed, a second screw can be inserted and a further check made. Should any recess be too deep, use a piece of cardboard to pack and adjust this. Before final hanging, it is advisable to remove the door and give the bottom edge one or two coats of paint.

Door stops are made of $\frac{1}{2}$in x 1in (13mm x 25mm) lumber nailed around the door frame $\frac{1}{16}$in (2mm) from the inner face of the closed door.

The door hardware can now be fitted. The height for this is optional, around 3ft (914mm) though this will look best if it lines up with any glazing bars in the doors or any other features such as the height of the sills.

The latch is fitted by squaring a line around the stile at the required height with try square and marking knife, and measuring the distance from the stile edge for the latch spindle. Bore a $\frac{1}{2}$in (13mm) hole for this. Now make a similar hole in the edge of the door, the size of which will depend on the size and shape of the latch barrel. Gauge the face of the barrel on to the face of the stile and drill a hole for a series of holes in line (Fig. 22); chop these out to accommodate the barrel. Fit this to the depth of the front plate and mark this on the stile. The depth of the plate can then be chiselled out. The latch spindle and plate and handles can then be fitted.

To find the position of the striking plate close the door and mark on the side the position of the latch tongue. Next put the plate in position on the door edge and close the door and mark on the side around this. Chop a small mortise in the centre to accommodate the tongue. Bend the lead-in part of the plate backward slightly and recess it as this will make the action smoother.

Glazing

You may wish to cut your own glass, but usually a glass supplier or handyman's shop can do this for you, $\frac{1}{4}$in (6mm) glass should be used.

There is a special variety of glass suitable for greenhouses which admits plenty of light yet keeps down the temperature, but this you may have to order in advance.

To cut glass, a steel glass wheel is satisfactory for most work and works out cheaper than the traditional glass cutter's diamond. You need a large, flat surface to cut glass. A felt-tipped pen can be used to mark guidelines on the glass. A long straightedge, a yardstick or a homemade tee square is needed to guide your cutter accurately along the lines.

First, clean the glass. To cut, use a firm stroke, holding the cutter vertically. Never backtrack, since the glass is unlikely to break along the cut line. After the surface has been scratched, put a strip of wood, or the straightedge, beneath the glass under the score line. Place your fingertips as closely as possible to the line and on both sides and press down slowly and firmly. You should get a clean break. If you have to trim surplus from the glass, scratch a further line and gently break off the waste in small bits

Gazebo

Fig 21. *The method used to mark and cut the recesses for the hinges.*

Fig 22. *The method used to cut the hole for the door latch.*

Fig 23. *The construction of the staging for the interior of the gazebo. Note that the sills and cladding have been omitted from the drawing for clarity.*

Fig 24 *Plan view of the gazebo showing the construction of the staging and position of the uprights which support it.*

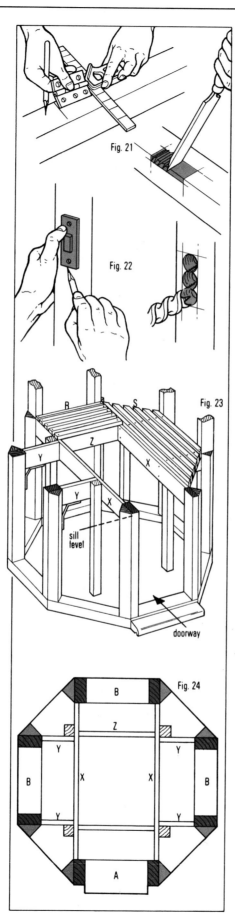

using pliers with emery cloth in the jaws.

Before glazing, apply a good primer to all lumber rabbets to prevent oil in the glazing compound from being sucked out. The glazing compound should be rolled in the hand until malleable. Use linseed oil to soften it, if necessary. Next line the rabbet with bedding compound by 'rolling' this in a thin strip from a ball in the hand. Now press the pane into place at the edges. Never press at the center. This will squeeze out surplus compound leaving a bed to a depth of $\frac{1}{16}$in to $\frac{1}{8}$in (2 to 3mm). Cut off the surplus with a putty or compound knife. Once glazing is completed the glazing beads can be tapped in to hold windows firmly.

Glazing presents little difficulty, although a few general points in glazing are: all four beads should be inserted before any of them is finally fixed in place; nails should not be inserted too near the corners.

Also, when smoothing glazing compound around newly glazed areas, it sometimes tends to stick to the putty knife. This can be avoided if the knife is kept moist with water, providing a smooth finish.

Benches and finishing

Benches can be fitted to choice in the greenhouse. The following is a suggested arrangement to meet average needs. This provides shelving racks at all levels, consisting of 1in x 4in (25mm x 102mm) framing with 1in x 2in (25mm x 51mm) decking (Figs. 23 and 24).

Framing is screwed by non-rusting metal brackets to the greenhouse frame uprights. Fixing of the cross members, staggered to facilitate this, is through the face of the cross bearers into the ends of those which butt to those at a right angle (Fig. 23). Non-rusting nails are used to make these fixings. Where heavy pots have to be carried on the benches, lumber center posts can be fitted.

The level of the shelf rail should be 1in (25mm) below that of the sill level so that the decking is at this height. Paving slabs can be laid inside, leaving earth at the sides for planting.

Finally the finish, inside and out, is a matter of choice. Hardwoods and cedar will not rot, but a preservative or clear varnish will prevent discoloration from weather. The lumber can be painted or finished with a polyurethane varnish. An attractive effect can be achieved by using a polyurethane woodstain varnish.

Index

If a page number in this index appears in *italics*, an illustration accompanies the text on that page.

Pictures supplied by
Peter Bell 137, 139
Harry Butler 45, 108, 112, 119
Roy Day 174, 181
Alan Duns 71
Nelson Hargreaves 8, 66/7, 70, 79, 80, 114,
  165, 166
John Hovell 164
Paul Kemble 154
R. Locke/Roy Flooks 169
Nigel Messett 51, 76, 85, 91, 105, 141, 145,
  148
Derek Metson 133
John Price 7, 16, 17, 18, 20, 22, 23, 25, 26,
  27, 29, 31, 33, 34, 36, 37, 43, 44
Richard Sharpe 55, 61
Tubby 46/7, 87
Colin Watmough 128/9

Illustrations
Tri-Art 99, 104, 150
All other diagrams by Bob Mathias